D1540148

Eeva Therman

Human Chromosomes
Structure, Behavior, Effects

With 55 Figures

122380

Springer-Verlag
New York Heidelberg Berlin

EEVA THERMAN
Department of Medical Genetics, University of Wisconsin, Madison, Wisconsin 53706, USA

Library of Congress Cataloging in Publication Data
Therman, Eeva.
  Human chromosomes.
  Includes bibliographies.
  1. Human genetics. 2. Human chromosomes.
3. Human chromosome abnormalities. I. Title.
[DNLM: 1. Chromosomes, Human. QH600 T411h]
QH431.T436   611'.01816   80-21278

All rights reserved.
No part of this book may be translated or reproduced in any form without written permission from Springer-Verlag.
The use of general descriptive names, trade names, trademarks, etc. in this publication, even if the former are not especially identified, is not to be taken as a sign that such names, as understood by the Trade Marks and Merchandise Marks Act, may accordingly be used freely by anyone.

© 1980 by Springer-Verlag New York Inc.
Printed in the United States of America

9 8 7 6 5 4 3 2 1

ISBN 0-387-90509-X
Springer-Verlag New York Heidelberg Berlin
ISBN 3-540-90509-X
Springer-Verlag Berlin Heidelberg New York

QH
431
T436

For *KLAUS PATAU*

6-22-82   adamar 18.80

Preface

This book provides an introduction to human cytogenetics. It is also suitable for use as a text in a general cytogenetics course, since the basic features of chromosome structure and behavior are shared by all eukaryotes. Because my own background includes plant and animal cytogenetics, many of the examples are taken from organisms other than man. Since the book is written from a cytogeneticist's point of view, human syndromes are described only as illustrations of the effects of abnormal chromosome constitutions on the phenotype. The selection of the phenomena to be discussed and of the photographs to illustrate them is, in many cases, subjective and arbitrary and is naturally influenced by my interests and the work done in our laboratory.

The approach to citations is the exact opposite of that usually used in scientific papers. Whenever possible, the latest and/or most comprehensive review has been cited, instead of the original publication. Thus the reader is encouraged to delve deeper into any question of interest to him or her.

I am greatly indebted to many colleagues for suggestions and criticism. However, my special thanks are due to Dr. JAMES F. CROW, Dr. TRAUTE M. SCHROEDER, and Dr. CARTER DENNISTON for their courage in reading the entire manuscript. I wish to express my gratitude also to the cytogeneticists and editors who have generously permitted the use of published and unpublished photographs.

Members of my laboratory have been very helpful in the process of putting this book together. Mrs. BARBARA SUSMAN has been involved in all the phases, from compiling reference lists to designing illustrations.

The photographic work has been done by Mr. WALTER KUGLER, Jr.; our excellent secretary, Ms. CHRISTINE ONAGA, has cheerfully typed and retyped the manuscript.

However, my most special thanks I wish to reserve for the editor of this book, Ms. MARY LOU MOTL. When my Finnish-English emerges from her competent hands, it resembles the text I would have liked to have written in the first place.

Madison, Wisconsin   EEVA THERMAN
September, 1980

Contents

VIII. Causes of Chromosome Breaks

IX. The Main Features of Meiosis

X. Details of Meiosis

XI. Meiotic Abnormalities

XII. Human Sex Chromosomes

XIII. Sex Chromosome Abnormalities

XVIII. Chromosomes and Cancer

XIX. Mapping of Human Chromosomes

Human Chromosomes

I
Past and Future of Human Cytogenetics

The Past of Human Cytogenetics

Before 1956 two "facts" were known about human cytogenetics. The human chromosome number was believed to be 48 and the XX-XY mechanism of sex determination was assumed to work in the same way as it does in *Drosophila*. Studies of the fruit fly had shown that the *ratio* of the number of X chromosomes to the number of sets of autosomes determines the sex of the organism. Both these fundamental notions about human chromosomes were eventually proved wrong.

The year 1956 is often given as the beginning of modern human cytogenetics, and indeed the discovery of Tjio and Levan (1956) that the human chromosome number is 46, instead of 48, was the starting point for subsequent spectacular developments in human chromosome studies. The difficulties of writing about even fairly recent history are well demonstrated by the very different accounts of this discovery related by the two participants themselves (Tjio, 1978; Levan, 1978).

The history of human cytogenetics has been reviewed at length several times, for instance, by Makino (1975) and recently by Hsu (1979). Hsu's delightful book relieves this author of the responsibility of giving a detailed description of the developments in the field; instead I will simply outline the major events in cytogenetics.

Hsu (1979) divides human cytogenetics conveniently into four eras: the dark ages before 1952, the hypotonic period from 1952 to about 1958, the trisomy period between 1959 and 1969, and the chromosome banding era that started in 1970 and still continues. In the following discussion only a few highlights of these stages will be recounted.

The Dark Ages

The difficulties faced by the early cytogeneticists are illustrated by a comparison of Fig. III.1 with the other photomicrographs of human chromosomes in this book. Despite the lack of clarity, the lymphocyte mitosis in Fig. III.1 shows the chromosomes considerably better than did the slides of paraffin-sectioned testes, stained with hematoxylin, that were used during the first quarter of this century. Of these studies only the paper by Painter (1923) is mentioned here, since it determined the ideas in this field for the next 33 years. Even though Painter's report that the human chromosome number was 48 was worded quite cautiously, the more often it was quoted, the more certain the finding seemed to become.

Despite the primitive techniques available, the groundwork for future studies was laid during the dark ages. The first satisfactory preparations of mammalian chromosomes were obtained by squashing ascites tumor cells of the mouse (Levan and Hauschka, 1952, 1953) and of the rat (cf. Makino, 1975). The first successful prefixation treatment with chemical substances was performed on mouse tumor cells by Bayreuther (1952); later colchicine or its derivatives were used.

During this era, mammalian tissue culture techniques were developed. Prefixation treatment with hypotonic salt solution, which swells the cells and thus separates the chromosomes, was a decisive improvement in cytological technique. The hypotonic treatment was launched by Hsu (1952), although other laboratories were experimenting with similar treatments at the same time.

The Hypotonic Era

The simultaneous use of a number of new techniques finally made it possible to establish the right chromosome number in man. They were the tissue culture and squash techniques combined with treatments with colchicine and hypotonic solution prior to fixation. Before the end of 1956, the finding of Tjio and Levan in embryonic lung cells was confirmed in human spermatocytes by Ford and Hamerton (1956) whose photomicrographs also showed that the X and Y chromosomes are attached end-to-end by their short arms in meiosis. During the hypotonic era the analysis of the human karyotype was also begun.

The Trisomy Period

The new techniques were soon applied to chromosome analyses of individuals who were mentally retarded or had other congenital anomalies or both. The first autosomal trisomy was described by Lejeune et al (1959) who found that mongolism (Down's syndrome) was caused by trisomy for one of the smallest human chromosomes. During the same year it was reported that Turner's syndrome was characterized by a 45,X chromosome constitution (Ford et al, 1959) and Klinefelter's syndrome by a 47,XXY chromosome complement (Jacobs and Strong, 1959). In addition, the first XXX woman was described (Jacobs et al, 1959). The observations on Turner's and Klinefelter's syndromes showed that the male sex in human beings is determined by the presence of the Y chromosome. Later it was established that the Y chromosome is effective in determining male sex even if it is combined with four X chromosomes; individuals with the XXXXY sex chromosome constitution are males, although abnormal.

The following year, D_1 trisomy (now known to be 13 trisomy) (Patau et al, 1960) and 18 trisomy (Edwards et al, 1960; Patau et al, 1960; Smith et al, 1960) were described. With these discoveries the viable autosomal trisomies seemed to be exhausted, although later the exceedingly rare 22 trisomy was found, and chromosome studies turned to structural aberrations and their phenotypic consequences.

These developments coincided with an important innovation in cell culture technique. Nowell (1960) and Moorhead et al (1960) launched the short-term culture technique using peripheral lymphocytes. The effectiveness of the technique was based on the mitosis-inducing ability of phytohemagglutinin. Such cultures, combined with the trick of drying the chromosomes directly on microscope slides (Rothfels and Siminovitch, 1958), are still the most important source of human and mammalian chromosomes.

Chromosome Banding Era

Despite all claims to the contrary, the chromosomes in groups B, C, D, F, and G could not be identified individually on morphological grounds (Patau, 1960); the numbers attached to the paired-off chromosomes in prebanding karyotypes represented sheer guesses. Although autoradiography had allowed the accurate identification of some chromosomes (cf. Patau, 1965), the degree of precision was increased by orders of magnitude with the introduction of chromosome banding techniques. In 1970, Caspersson et al applied fluorescence microscopy, which they had originally used to study plant chromosomes, to the analysis of the human karyotype. They discovered that the chromosomes consist of differentially fluorescent cross bands of various lengths. Careful study of these bands made possible the identification of all human chromosomes. This discovery was followed by a flood of different banding techniques that utilize either fluorescent dyes or the Giemsa stain. The banding of prophase chromosomes makes it possible to determine chromosome segments and breakpoints even more accurately (Yunis, 1976).

Another milestone was the discovery that chromosomes that incorporate bromodeoxyuridine (BrdU) instead of thymidine have different staining properties. This phenomenon has been successfully used to reveal the late-replicating chromosomes and chromosome segments (Latt, 1974). It also provides the basis for the study of sister chromatid exchanges (Latt, 1973).

It is much more difficult to obtain satisfactory chromosome preparations of the male meiosis—not to mention the female meiosis—in man than, for instance, in the mouse. But lately these difficulties have to some extent been overcome. The early stages of meiosis have been analyzed successfully in the oocytes (e.g., Therman and Sarto, 1977; Hultén et al, 1978), whereas work on the spermatocytes has yielded clear photomicrographs of the later stages (e.g., Stahl et al, 1973).

Human Sex Chromosomes

Throughout Hsu's (1979) four eras the understanding of the function and behavior of the mammalian sex chromosomes increased steadily. One of the first important observations was that the neural nuclei of the female cat had a condensed body, missing in the male nuclei (Barr and Bertram, 1949). This body has been called sex chromatin, the Barr body, or X chromatin.

The single-active X hypothesis of Lyon (1961; cf. Russell, 1961) had a decisive influence on the entire field of mammalian sex chromosome studies. According to the Lyon hypothesis, as it is called, one X chromosome in mammalian female cells is inactivated at an early embryonic stage. The original choice of which X is inactivated is random, but in all the descendants of a particular cell the same X remains inactive. If a cell has more than two X chromosomes, all but one of them are turned off. This mechanism provides dosage compensation for X-linked genes because each cell, male or female, has only one X chromosome that is transcribed. The Barr body is formed by the inactive X chromosome (Ohno and Cattanach, 1962), which is out of step with the active X chromosome during the cell cycle.

One of the highlights in the study of mammalian sex determination is the recent discovery that the primary sex determination of the Y chromosome is mediated through the H-Y antigen, which is a plasma membrane protein (cf. Ohno, 1979). This antigen induces the development of testicular tissue, which in turn determines secondary sex development through its production of androgen.

Evolution of Human Chromosomes

The Phylogeny of Human Chromosomes, as Seuánez (1979) calls his book on the subject, has been studied intensively in recent years. A comparison of the chromosomes of man with those of his closest relatives—the chimpanzee, gorilla, and orangutan—shows that 99 percent of the chromosome bands are shared by the four genera. The most prominent differences in banding patterns occur in the heterochromatic regions (cf. Seuánez, 1979). Surprisingly it appears that man's closest living relative is the gorilla and not the chimpanzee, as has been believed until now.

The similarity of the chromosome banding pattern in all four genera demonstrates that most of the individual bands have retained their identity for more than 20 million years, and many of them for considerably longer. A number of chromosomes in man and the great apes are identical. The most conservative of these chromosomes is the X, which has not changed in morphology, at least between the monkey and man. Its gene content is assumed to have remained the same throughout mammalian development, or for some 125 million years (cf. Ohno, 1967; Seuánez, 1979). The comparison of the chromosomes of man and his relatives is now under way on the molecular level too (cf. Jones, 1977).

Nomenclature of Human Chromosomes

As the number of laboratories involved in analysis of human chromosomes multiplied, so did the systems of chromosome designation. In an effort to create order in this threatening chaos, four conferences on chromosome nomenclature were held: in Denver (1960), London (1963), Chicago (1966), and Paris (1971) (cf. Makino, 1975). The recommendations of the latest conference (Paris Conference, 1971) which included the designations for the chromosome bands, are now in use (Chapter V).

The Future of Human Cytogenetics

The expansion of the science of human cytogenetics in somewhat more than 20 years is little short of miraculous; by now, from the viewpoint of cytogenetics, man is by far the most extensively studied organism. During its early stages, human cytogenetics was a more-or-less applied science: phenomena previously described in plants and animals were now being observed in man. However, human cytogenetics has come of age, and advances in this field have inspired studies in other branches of human genetics. Indeed it is the coordination of different approaches that has led to the most interesting results in this field. During the "dark ages", human cytogeneticists borrowed techniques from plant and animal studies. Now the opposite is often true, and both animal and plant chromosome studies owe a debt to the work done on humans.

Predictions of future developments in a scientific field can only be based on its present state. However, just as unexpected findings in the past have changed the course of events, they will undoubtedly do so in the future. In the following discussion, those approaches to human cytogenetics that seem most promising to this author are briefly reviewed.

Structure of the Eukaryotic Chromosome

Banding techniques added a new dimension to the understanding of the longitudinal differentiation of human chromosomes (Chapter V and VI). The quinacrine-bright bands seem to contain more heterochromatin, whereas the dark bands are more gene-rich. The main locations of constitutive heterochromatin have been determined, and different types of heterochromatin distinguished on the basis of their staining properties. Immunofluorescent stains that are specific for different chromosome constituents seem to offer a promising approach.

Only a few years ago very little information existed on the fine structure of the eukaryotic chromosome. There was an almost total gap between our knowledge of chromosomes, as seen in the light microscope, and what was known about DNA in vitro. This gap is now gradually being bridged, and in the not too distant future an understanding of how the chromosome is built out of its constituents ought to emerge.

A chromatid consists of one double helix of DNA. Its structure is determined by the base ratios and the arrangement of the bases. This structure is reflected in the distribution of the different classes of histones and other proteins along the chromosomes; the location of the proteins in turn determines the visible banding pattern.

Chromosomal DNA has been studied in vitro with different techniques. Based on the speed of renaturation of isolated, sheared DNA, it has been divided into fractions containing highly repetitive, intermediately repetitive, or unique sequences. Another approach to the analysis of DNA is the fractionation of native double-stranded DNA by cesium salt-density gradients. This technique shows that minor components, so-called satellite DNAs, differ in their buoyant densities from the main bulk of DNA. Hybridization of the various isolated DNA fractions (or the RNAs coded from them) on chromosomes demonstrates that repetitive sequences, as well as various satellite DNAs, are mainly localized in the bands representing constitutive heterochromatin. However, the exact relationships of the DNA fractions, isolated by different methods to each other and to the chromosome bands, are still largely unknown.

Another unanswered question is: What role does the ubiquitous, but apparently inert, heterochromatin play in the cell? Further, what is the function of the great amount of DNA in eukaryotic chromosomes, which does not represent constitutive heterochromatin but is not being transcribed either? The base sequencing of DNA, now under way, may provide answers to some of these questions.

Mapping of Human Chromosomes

The development of an accurate gene map is one of the main goals of cytogenetic studies in any organism. The localization of genes to specific chromosomes and chromosome segments is one of the most rapidly advancing branches of human cytogenetics (Chapter XIX). The methodology includes linkage studies, the use of marker chromosomes in family studies, in vitro fusion of human cells with cells of other species, and direct hybridization of DNA and RNA on chromosomes. The positions of the repeated genes, such as those coding for ribosomal proteins or histones, have been determined by hybridization. Soon it should even be possible to determine the positions of single genes by direct hybridization. In evolutionary studies, comparative gene mapping in related organisms seems to have an almost unlimited future.

Chromosomes and Cancer

The analysis of malignant cells has been the subject of an enormous amount of work, but so far many of the results remain ambiguous (Chapter XVIII). The main reason for this fact, apart from technical difficulties, is that it is impossible to distinguish the primary chromosome change—if there is such a thing—from those that arise during the development of the cancer. However, with improved cytological techniques , some principles have begun to emerge from the apparent chaos. On the one hand, so far a definite chromosomal cause has been established in very few types of cancer, although it is expected that their number will increase. On the other, it is observed that the secondary chromosome aberrations that occur during the development of malignant disease are highly *nonrandom*. In other words, many types of tumors and leukemias are characterized by what appears to be a predetermined sequence of chromosomal changes.

Mutagenesis Studies

Chromosome breakage has been used for a long time as an indicator of the mutagenic effects of various agents (Chapter VIII). The resolution of such studies has been greatly increased by chromosome banding. Chromosome breaks occur nonrandomly along the chromosomes; the breaks are mainly localized in the Q-dark bands, some of which constitute veritable "hot spots."

The introduction of sister chromatid exchanges as a test system has caused a true revolution in mutagenesis testing. Sister chromatid exchanges are not only a much more sensitive indicator of mutagenic activity than chromosome breaks but also considerably easier to score unambiguously.

Cytology of Development

A branch of mammalian cytogenetics that, in my opinion, is just beginning is the analysis of what happens in the nuclei during development (Chapter IV). Important results in developmental cytology were achieved in plants and insects during the 1930s and 1940s. In mammals, however, mitotic modifications and their effects have been studied almost exclusively in liver, bone marrow, and malignant cells. In addition to spectrophotometry, in use for 30 years, new techniques, such as flow cytometry, analysis of prematurely condensed chromosomes in cells fused at different stages of the cell cycle, and differential staining techniques are being applied to the analysis of interphase nuclei. Studies like these may give us an idea of the role that nuclear changes play in differentiation.

Clinical Cytogenetics

The cooperation between basic scientists and cytogeneticists in laboratories where human chromosomes are studied for medical purposes has been very fruitful. Most advances in human cytogenetics have also led, on the one hand, to improved clinical applications. On the other, the enormous number of normal and abnormal persons whose chromosomes have been analyzed provides valuable material for many types of theoretical studies (Chapter XIII–XVII).

Banding techniques have allowed the identification of new chromosomal syndromes, for instance, 22 trisomy and 21 monosomy, as well as conditions caused by trisomy mosaicism for different chromosomes. In addition an apparently unlimited variety of partial trisomy and monosomy syndromes has emerged (Chapter XV). The accuracy of breakpoint determinations in structurally abnormal chromosomes will naturally increase when the prophase banding technique comes into general use.

It is expected that these developments will lead to specific risk figures for different types of chromosome translocations (and for other chromosome aberrations), which so far have been pooled for purposes of genetic counseling (Chapters XVI and XVII). In addition, the possible genetic risks posed by extreme heterochromatic variants as well as by position effects, which have been established for the X chromosome and assumed for the autosomes, are now under study.

References

Barr ML, Bertram EG (1949) A morphological distinction between neurons of the male and female, and the behavior of the nuclear satellite during accelerated nucleoprotein synthesis. Nature 163: 676–677

Bayreuther K (1952) Der Chromosomenbestand des Ehrlich-Ascites-Tumors der Maus. Naturforsch 7: 554–557

Caspersson T, Zech L, Johansson C (1970) Differential banding of alkylating fluorochromes in human chromosomes. Exp Cell Res 60: 315–319

Edwards JH, Harnden DG, Cameron AH, et al (1960) A new trisomic syndrome. Lancet i: 787–790

Ford CE, Hamerton JL (1956) The chromosomes of man. Nature 178: 1020–1023

Ford CE, Jones KW, Polani PE, et al (1959) A sex-chromosome anomaly in a case of gonadal dysgenesis (Turner's syndrome). Lancet i: 711–713.

Hsu TC (1952) Mammalian chromosomes in vitro. I. The karyotype of man. J Hered 43: 167–172

Hsu TC (1979) Human and mammalian cytogenetics. An historical perspective. Springer, Heidelberg

Hultén M, Luciani JM, Kirton V, et al (1978) The use and limitations of chiasma scoring with reference to human genetic mapping. Cytogenet Cell Genet 22: 37–58

Jacobs PA, Baikie AG, Court Brown WM, et al (1959) Evidence for the existence of the human "super female." Lancet ii: 423–425

Jacobs PA, Strong JA (1959) A case of human intersexuality having a possible XXY sex-determining mechanism. Nature 183: 302–303

Jones KW (1977) Repetitive DNA and primate evolution. In: Yunis JJ (ed) Molecular structure of human chromosomes. Academic, New York, pp 295–326

Latt SA (1973) Microfluorometric detection of deoxyribonucleic acid replication in human metaphase chromosomes. Proc Natl Acad Sci USA 70: 3395–3399

Latt SA (1974) Microfluorometric analysis of DNA replication in human X chromosomes. Exp Cell Res 86: 412–415

Lejeune J, Gautier M, Turpin R (1959) Etude des chromosomes somatiques de neuf enfants mongoliens. Compt Rend 248: 1721–1722

Levan A (1978) The background to the determination of the human chromosome number. Am J Obstet Gynecol 130: 725–726

Levan A, Hauschka TS (1953) Endomitotic reduplication mechanisms in ascites tumors of the mouse. J Natl Cancer Inst 14: 1–43

Levan A, Hauschka TS (1953) Endomitotic reduplication mechanisms in ascites tumors of the mouse. J Natl Cancer Inst. 14: 1–43

Lyon MF (1961) Gene action in the X-chromosome of the mouse. Nature 190: 372–373

Makino S (1975) Human chromosomes. Tokyo, Igaku Shoin

Moorhead PS, Nowell PC, Mellman WJ, et al (1960) Chromosome preparations of leucocytes cultured from human peripheral blood. Exp Cell Res 20: 613–616

Nowell PC (1960) Phytohemagglutinin: an initiator of mitosis in cultures of normal human leukocytes. Cancer Res 20: 462–466

Ohno S (1967) Sex chromosomes and sex-linked genes. Springer, Heidelberg

Ohno S (1979) Major sex-determining genes. Springer, Heidelberg

Ohno S, Cattanach BM (1962) Cytological study of an X-autosome translocation in *Mus musculus*. Cytogenetics 1: 129–140

Painter TS (1923) Studies in mammalian spermatogenesis. II. The spermatogenesis of man. J Exp Zool 37: 291–336

Paris Conference (1971) Standardization in human cytogenetics. Birth defects: original article series, VIII: 7. New York, The National Foundation, 1972

Patau K (1960) The identification of individual chromosomes, especially in man. Am J Hum Genet 12: 250–276

Patau K (1965) Identification of chromosomes. In: Yunis JJ (ed) Human chromosome methodology. Academic, New York, pp 155–186

Patau K, Smith DW, Therman E, et al (1960) Multiple congenital anomaly caused by an extra autosome. Lancet i: 790–793

Rothfels KH, Siminovitch L (1958) An air-drying technique for flattening chromosomes in mammalian cells grown in vitro. Stain Tech 33: 73–77

Russell LB (1961) Genetics of mammalian sex chromosomes. Science 133: 1795–1803

Seuánez HN (1979) The phylogeny of human chromosomes. Springer, Heidelberg

Smith DW, Patau K, Therman E, et al (1960) A new autosomal trisomy syndrome: multiple congenital anomalies caused by an extra chromosome. J Pediatr 57: 338–345

Stahl A, Luciani JM, Devictor-Vuillet M (1973) Etude chromosomique de la meiose. In: Boué A, Thibault C (eds) Les accidents chromosomiques de la reproduction. Paris, I.N.S.E.R.M., Centre International de l'Enfance, pp 197–218

Therman E, Sarto GE (1977) Premeiotic and early meiotic stages in the pollen mother cells of Eremurus and in human embryonic oocytes. Hum Genet 35: 137–151

Tjio JH (1978) The chromosome number of man. Am J Obstet Gynecol 130: 723–724

Tjio JH, Levan A (1956) The chromosome number in man. Hereditas 42: 1–6

Yunis JJ (1976) High resolution of human chromosomes. Science 191: 1268–1270

II

Structure of the Eukaryotic Chromosome and the Karyotype

Metaphase Chromosome

Higher organisms are *eukaryotes* in contrast to bacteria and phages, which are *prokaryotes*. The eukaryotic chromosome is a complicated structure that, in addition to DNA, contains several different types of proteins. A prokaryotic chromosome consists of naked DNA or in some phages, naked RNA. The chromosomes of higher organisms are studied most frequently at mitotic metaphase. This is the stage at which the chromosomes reach their greatest condensation, and this natural condition is increased by a prefixation treatment with various drugs, for example, colchicine. During mitotic metaphase, the condensed chromosomes appear in identifiable shapes characteristic of the karyotype of the species being studied.

Primary Constriction

A typical metaphase chromosome consists of two arms separated by a *primary constriction*, which is made more clearly visible by treatment with colchicine. This constriction marks the location of the *centromere* or *spindle attachment* (sometimes called the kinetochore), which is essential for the normal movements of the chromosomes in relation to the spindle. A chromosome without a centromere is an *acentric* fragment and either is lost or drifts passively along when the other chromatids move to the poles, by the action of the spindle, during anaphase.

In the plant genera of rushes (*Juncus*) and sedges (*Carex*) as well as in certain insects and the scorpions, the centromere is diffuse. This means that a small chromosomal fragment, even when separated from the rest of the chromosome, acts as a complete chromosome and displays normal anaphase movements.

A metaphase chromosome consists of two *sister chromatids* that separate in mitotic anaphase. The genetic constituent of a chromatid is a double helix of DNA. Cytologists have long disputed whether a chromatid consists of one double helix of DNA or several parallel ones. It has been assumed that the latter situation prevails, especially in organisms with large chromosomes. However, recent observations show convincingly that each chromatid of a eukaryotic chromosome contains one double helix of DNA continuous from one end of the chromosome to the other. In other words, a chromatid is a *unineme* structure. Certain specialized tissues, such as the endosperm in plant seeds wherein the chromatids consist of more than one DNA strand, provide exceptions to this rule. How widespread such *multineme* chromosomes are is not presently clear.

Through a series of coils within coils, the chromosome strands, or *chromonemas*, shorten greatly between interphase and metaphase (see Fig. 1 in Ris and Korenberg, 1979). The largest coil is sometimes visible in the light microscope (Fig. II.1), as in a mouse cancer chromosome treated with 1-methyl-2-benzyl hydrazine, which displays segments with distinct coiling interspersed with condensed regions (Therman, 1972). In addition to DNA, eukaryotic chromosomes contain nonhistone proteins together with five types of histones (cf. Ris and Korenberg, 1979), all of which seem to play an important role in the condensation and coiling of chromosomes.

Secondary Constrictions

In addition to the primary constriction, a chromosome may contain a *secondary constriction,* which appears as a constriction or an unstained gap in the chromosome. Usually a secondary constriction contains a *nucleolar organizer* (Fig. II.2). Secondary constrictions may be situated anywhere along the chromosome. However, they are most often near an end, separating a small segment, a *satellite,* from the main body of the chromosome. In such cases the secondary constriction is called a *satellite stalk*.

Characterization of Metaphase Chromosomes

A metaphase chromosome is identified morphologically both by its total length and by the position of the centromere, which determine the relative lengths of its arms. A secondary constriction also helps in the identification of a particular chromosome. Nowadays chromosomes are usually identified with banding techniques (Chapters V and VI).

A chromosome in which the centromere is more or less in the middle is called *metacentric*. In an *acrocentric* chromosome, the arms are

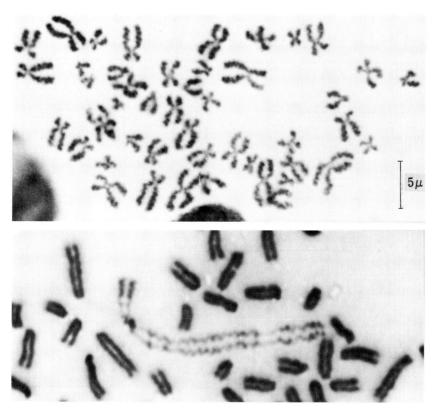

Fig. II.1 Spiralization of metaphase chromosomes. Top: human lymphocyte. Bottom: mouse cancer cell treated with 1-methyl-2-benzyl hydrazine (Therman, 1972).

markedly unequal. Chromosomes intermediate between these two are *submetacentric* and *subtelocentric*. A chromosome in which the centromere is at the very end is *telocentric*. Truly telocentric chromosomes are almost nonexistent in natural populations, but they have been found in some very unusual individuals, including man.

In human chromosomes, the short arm is designated *p* (petit) and the long arm is *q* (the next letter in the alphabet). Chromosomes are usually characterized by one of two parameters. The *arm ratio* (q/p) is the length of the long arm divided by that of the short arm. The *centromere index* expresses the percentage of the short arm in terms of the total chromosome length $\{[p/(p+q) \cdot 100\%\}$. The length of a particular chromosome relative to the others in the same plate, together with the arm ratio or centromere index, is sometimes sufficient to permit identification of the chromosome.

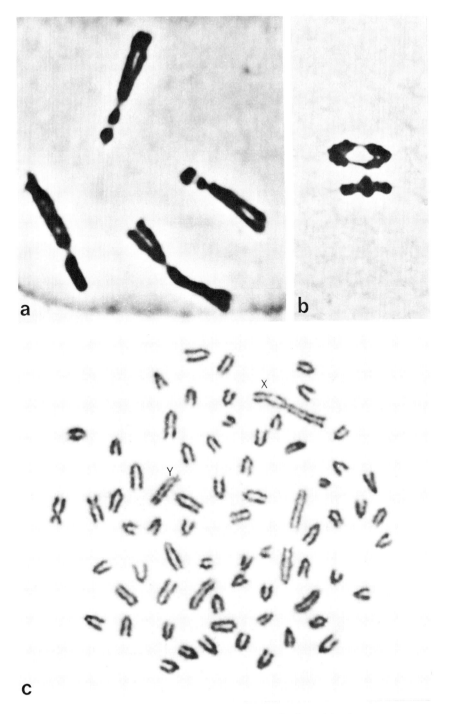

Fig. II.2. (a) Metaphase of *Haplopappus* with two subtelocentric chromosomes (nucleolar constrictions in the short arms) and two submetacentric chromosomes; (b) two bivalents in I meiotic metaphase of *Haplopappus* (a and b courtesy of RC Jackson); (c) metaphase of male reindeer (2*n* = 76) showing X and Y chromosomes.

Chromosome Number

The diploid chromosome number of an organism, usually determined by counting the chromosomes in dividing somatic cells, is indicated by the symbol $2n$. The gametes have one-half the diploid number of chromosomes (a haploid set), which is indicated by n.

The chromosome numbers vary greatly between and within groups of organisms. The chromosome numbers show no clear trend of becoming either higher or lower during evolution. A haploid number $n = 2$ is extremely rare, the best known example being the compositous plant, *Haplopappus* (Fig. II.2), which grows in the southwest United States. In mammals, the lowest haploid number $n = 3$ has been found in the muntjac, a small Indian deer. The largest haploid number reported in a higher organism is about $n = 630$ in a fern (*Ophioglossum reticulatum*). In mammals the highest haploid number $n = 46$ has been observed in a rodent (*Auctomys leander*). The chromosome complements established for mammalian species have been reviewed by Hsu and Benirschke (1967–1977).

The haploid numbers of most organisms are between 6 and 25. In Fig. II.2 two widely different chromosome constitutions are illustrated. In *Haplopappus* ($2n = 4$) one pair of chromosomes is submetacentric; the other is subtelocentric and has a nucleolar constriction in the short arm. The reindeer has, for a mammal, a relatively high chromosome number ($2n = 76$). Apart from a few subtelocentric chromosome pairs, most reindeer chromosomes are acrocentric. The metacentric X and the acrocentric Y chromosome are clearly distinguishable from the autosomes.

Chromosome Size

Chromosome size also varies widely in different organisms, ranging from a fraction of a micron (μ) in length, which is at the limit of resolution of the light microscope, to more than 30 μ. Very small chromosomes are found, for instance, in the fungi and green algae, whereas the largest ones have been observed in some amphibians and liliaceous plants. Most grasshoppers also have large chromosomes.

Although the chromosome complements of different organisms tend to be more similar the more closely related they are, there are striking exceptions to this rule. One of the classic examples of a great size difference in the chromosomes of two related species with the same diploid number ($2n = 12$) is provided by the leguminous plants *Lotus tenuis*, in which the mean length of the chromosomes is 1.8 μ, and *Vicia faba*, in which the corresponding value is 14.0 μ (cf. Stebbins, 1971).

In general, higher organisms tend to have larger chromosomes than do lower ones. Exceptions to this rule are numerous, however, both among unicellular and multicellular organisms.

The chromosomes within the same chromosome complement usually fall into a fairly limited size range; in other words, they all tend to be either large or small. Yet in a few instances in some animal groups, such as birds and lizards, the chromosome complements consist of a number of large chromosomes and a higher number of very small microchromosomes. In metaphase the small chromosomes usually lie in the middle of the plate, whereas the large ones form a circle around them.

Shape of Chromosomes

In addition to the number and size of the chromosomes, the chromosome complement of a species is characterized by the shape of the chromosomes. They may all be of one type or a combination of different types. In the mouse, for instance, all the chromosomes are acrocentric (Fig. VIII.1); in man the chromosomes range from metacentric to acrocentric (Figs. II.1 and 3). Within a group, the more highly developed species tend to have more asymmetrical (the two arms are unequal) chromosomes, which have evolved from species with more metacentric complements (cf. Stebbins, 1971).

Fig. II.3 I. Normal human male karyotype from a lymphocyte. II. Chromosomes 1 and 9 showing secondary constrictions. III. Chromosomes 1, 9, and 16 showing fuzzy regions. IV. G and Y chromosomes from father and son (orcein staining).

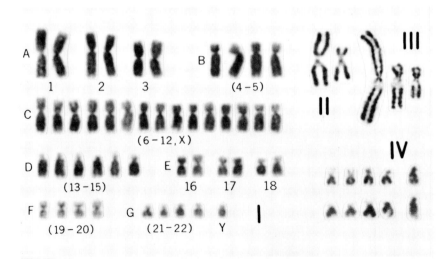

DNA Content of Nuclei

The DNA content of a nucleus is naturally determined by the number and size of the chromosomes of the organism. In the animal kingdom, the values range from 168.0 picogram (pg) per diploid nucleus in the salamander *Amphiuma,* which has very large chromosomes, to 0.2 pg in Drosophila, a 1000-fold difference. In plants the differences are almost as great, with values ranging from 196.7 pg in the liliaceous plant *Fritillaria* to 1.4 pg in the flax (*Linum usitatissimum*) (cf. Rees and Jones, 1977). In man, the DNA content of a diploid nucleus is 6.4 pg (cf. Rees and Jones, 1973). The values in other mammals deviate surprisingly little from this, especially when we consider the variation shown by their chromosome numbers.

Human Chromosome Complement

Man has 44 autosomes (nonsex chromosomes) and two sex chromosomes (two X chromosomes in the female and one X and one Y chromosome in the male). Human chromosomes range in size from somewhat larger than 5 μ to less than 1 μ; the range, however, varies between cells. With respect to both chromosome number and size, man stands in the middle range of higher organisms.

The term *karyotype* describes a display of the chromosomes of an organism such that they are lined up, starting with the largest and continuing in order of diminishing size. The convention dictates that the shorter chromosome arm points to the top of the page. An *idiogram* is a diagrammatic karyotype based on chromosome measurements in many cells. Figure II.3 illustrates the karyotype of a normal human male. Morphological identification is based on the relative size of the chromosomes and their arm ratio. According to these criteria, 1, 2, 3, 16, 17, 18, and the Y can be individually distinguished. Chromosome 9 (C′ in prebanding karyotypes) sometimes shows a fuzzy region next to the centromere on the long arm and can be identified on that basis. The rest of the chromosomes can be classifed only as belonging to the groups B, C (which contains the X chromosome), D, F, and G. Chromosomes 1 and 3 are typically metacentric; chromosome 2 is on the borderline between metacentric and submetacentric; the B chromosomes represent the subtelocentric type. The D and G chromosomes are acrocentric and usually display satellites on the short arms. The size of both short arms and satellites varies in different persons. Sometimes satellites are so small that they are practically invisible under the light microscope.

With the banding techniques now available, all the human chromosomes can be identified. However, even in the prebanding era, attempts were made to distinguish the chromosomes by measuring them and pairing off those that were most similar in size. However, even identifiable homologous chromosomes, say two number 1 chromosomes in the same metaphase plate, may show considerable difference in length. The coefficient of variation may be 5 percent or more. Since the real size difference between nonhomologous B, C, or D chromosomes is *less than* 5 percent (see Table II.1), pairing off by measurement alone clearly cannot lead to the right identification of chromosomes, except sometimes by chance (Patau, 1965). Despite this fact, attempts have been made to identify the X chromosome by pairing off the C chromosomes in a male. Those who started with the longest pair claimed that X was the shortest C chromosome; and those who started from the shortest chromosomes found that the partnerless X was the largest of the C group chromosomes. In reality it is the third largest (Table II.1). Obviously the pairing-off method has no basis for success. It is equally useless in identifying members of the other chromosome groups. Consequently most karyotype numbers in prebanding days represent sheer guesses and should be thus interpreted (cf. Patau, 1965).

Table II.1. Relative Lengths of Human Chromosomes.[a]

Chromosome Group	No.	Average length (in % of autosomal genome) of: Long arm	Short arm	Chromosome Group	No.	Average length (in % of autosomal genome) of: Long arm	Short arm
A	1	4.68	4.57		13	3.29	—
	2	5.28	3.35	D	14	3.12	—
	3	3.80	3.32		15	2.89	—
B	4	4.85	1.84		16	1.93	1.34
	5	4.66	1.75	E	17	2.07	0.96
	6	3.87	2.36		18	2.04	0.76
	7	3.54	2.04	F	19	1.32	1.11
	8	3.45	1.63		20	1.30	1.05
C	9	3.23	1.72	G	21	1.26	—
	10	3.22	1.54		22	1.38	—
	11	2.90	1.88	Total autosomes		100.00	
	12	3.38	1.32				
					X	3.26	2.02
					Y	1.64	—

[a]Revised by Patau (unpublished).

— Short arms not measurable.

In addition to being characterized by their size and arm ratio, certain chromosomes can be distinguished by fuzzy regions next to the centromere. These have also been called secondary constrictions. Such regions are now known to contain constitutive heterochromatin (Chapter VI). The fuzzy regions occur in chromosome 1 (by definition lq is the arm bearing this region), 9, and 16 (Fig. II.3). The Y chromosome is distinguishable from the G group by the lack of satellites on its short arm, a fuzzy distal end of the long arm—now known to be heterochromatic—and a secondary constriction sometimes visible in the middle of the long arm. Whereas in colchicine-treated metaphases the chromatids of the G chromosomes are spread out, the chromatids of Yq usually stick together.

The reader may wonder why the prebanding human karyotype is discussed at such length when the banded chromosomes yield so much more information. The reason is that a cytogeneticist must be able to read the older literature intelligently. To do this, one must have an understanding of what could and could not be seen with techniques available at different times.

Autoradiography, which is based on the fact that different chromosomes and chromosome segments replicate at different times during the synthetic period in interphase, provided a step forward in chromosome identification. This technique allows the observer to distinguish the inactive X chromosome in the female. The phenomenon of X inactivation will be discussed in Chapter XII. Also, chromosomes 4 and 5 can be distinguished, as can 13, 14, and 15, and 21 and 22 (cf. Patau, 1965). However, autoradiography is a laborious process and, where chromosome identification is concerned, has been superseded by the banding techniques. It will therefore not be discussed further in this connection.

Banded Human Karyotype

Although the banding techniques and their use are discussed in detail in Chapters V and VI, a G-banded human karyotype is presented here (Fig. II.4). The chromosomes are numbered according to Caspersson et al (1971; see also Paris Conference, 1971), who first published a banded human karyotype. The chromosomes are arranged numerically according to length with one exception: chromosome 22 is actually longer than 21. Since the chromosome that, in the trisomic state, causes mongolism (Down's syndrome) has long been called 21 in the literature, it was thought impractical and confusing to reverse the numbers. Chromosome 21 is, therefore, defined through the syndrome it causes. As seen in Fig. II.4, each human chromosome can be distinguished by its banding pattern. Most of the individual arms can also be unambiguously identified.

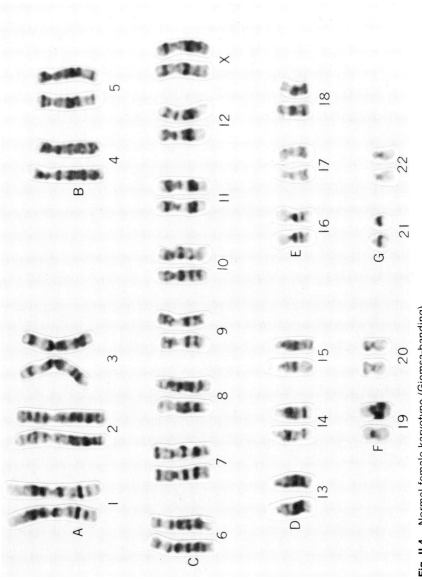

Fig. II.4. Normal female karyotype (Giemsa-banding).

In Table II.1 the lengths of human chromosome arms expressed as a percentage of the length of the haploid autosomal set or genome are given. The values are averages based on the compilation of a large number of measurements by different authors, with certain systematic errors corrected (Patau, unpublished). The lengths of short arms of the acrocentric chromosomes are not given, because first, they vary between individuals and second, anything under 1 μ (which would be about 1 percent of haploid karyotypes) is only about twice the wavelength of visible light. Because of this fact, the measurement of the much smaller short arms would be of dubious value.

Euploid Chromosome Changes

An interesting question is: How have the widely different karyotypes diverged from each other during evolution? All evolution is based on changes that occur in genetic material and upon which natural selection acts. The genetic material may undergo qualitative gene changes, or gene mutations. However, the evolution of karyotypes mainly depends on changes in quantity or arrangement of genes. These larger changes constitute the so-called chromosome mutations.

Changes in the number of whole sets of chromosomes, or *euploid* changes, usually lead to *polyploidy. Haploidy,* as a rule, is not a viable condition. Polyploidy can be inferred when the chromosome number of an organism is a multiple of the chromosome number of a related species. A chromosome complement containing three haploid sets is called *triploid (3n).* Four sets constitute a *tetraploid* chromosome complement *(4n).* It is estimated that of the higher plants, about one-half have chromosome numbers that are multiples of those of other related species. In certain plant families, such as grasses (*Graminae*) or roses (*Rosaceae,*) three-fourths of the species are polyploid. For example, in the genus *Rubus* of the rose family, the following multiples of the basic number 7 are found: 14, 21, 28, 35, 42, 49, 56, 63, and 84. In the compositous genus *Chrysanthemum* (n = 9), the diploid numbers range from 18 to 198 (cf. Darlington and Wylie, 1955).

In plant evolution, polyploidy is often combined with hybridization between different species. When the chromosome number of a hybrid is duplicated, we may in one step see the emergence of a new species combining the diploid complements of the parents. A well-known cultivated plant of this type is *Triticale,* a hybrid of wheat and rye with the chromosome complements of both species (*2n* = 56). However, one of the parents, the cultivated wheat, is a hexaploid with six sets of 7 chromosomes (one of the basic numbers in the grasses). Rye, on the other hand, is a diploid with *2n* = 14 chromosomes.

Polyploidy, which has been very important in plant evolution, seems to have played almost no role in animal evolution. Polyploidy in animals upsets the sex determination mechanism, and this has generally been assumed to block the successful establishment of polyploidy in them. Indeed most polyploid animal species are *parthenogenetic;* they produce offspring *without* meiosis or fertilization. As a rule, parthenogenetic offspring are uniform and genetically identical to the parent. Because parthenogenesis takes place without meiosis or fertilization, the results of the process naturally lack the variability created by meiotic crossing-over and segregation.

From an evolutionary point of view, parthenogenetic species have reached a dead end, no matter how successful they may be under specific prevailing conditions. Polyploid parthenogenetic forms are found in shrimp, earthworms, and some insects (cf. Darlington, 1958). In plants, asexual reproduction through rhizomes, bulbs, runners, and other organs is common, especially in polyploid species.

Apomictic seed formation corresponds to parthenogenesis in animals. However, all organisms reproducing asexually can be said to have sold their future for present advantage (cf. Darlington, 1958).

Aneuploid Chromosome Changes

Chromosome mutations also include changes in the number of individual chromosomes, as opposed to sets of chromosomes. These changes in the number of normal chromosomes are referred to as *aneuploidy.* The absence of one chromosome from the diploid complement is *monosomy;* the presence of an extra one is *trisomy.* A chromosome complement with two extra chromosomes is *tetrasomic.*

Structural Changes in Chromosomes

The third type of chromosome mutation occurs as a result of chromosome breakage and rejoining in such a way that the chromosomes are structurally reorganized. The details of chromosome rearrangements are discussed in Chapter VII. Such changes may sometimes result in an increase or decrease in chromosome segments as well as in changes of their order.

Various changes in chromosome number and structure have created the multitude of karyotypes observed in plants and animals. However, whatever selective advantage—few or numerous, large or small, symmetric or asymmetric—chromosomes bestow on a given species is a much discussed but still unresolved question (cf. Stebbins, 1971).

References

Caspersson T, Lomakka G, Zech L (1971) 24 fluorescence patterns of human metaphase chromosomes—distinguishing characters and variability. Hereditas 67: 89–102

Darlington CD (1958) The evolution of genetic systems. Basic Books, New York

Darlington CD, Wylie AP (1955) Chromosome atlas of flowering plants. Hafner, New York

Hsu TC, Benirschke K (1966–1977) An atlas of mammalian chromosomes. Springer, Heidelberg, Vol 1–10

Paris Conference (1971) Standardization in human cytogenetics. Birth defects: original article series, VIII: 7. New York: The National Foundation, 1972

Patau K (1965) Identification of chromosomes. In: Yunis JJ (ed) Human chromosome methodology. Academic, New York pp 155–186

Rees H, Jones RN (1977) Chromosome genetics. University Park Press, Baltimore

Ris H, Korenberg JR (1979) Chromosome structure and levels of chromosome organization. In: Goldstein L, Prescott DM (eds) Cell biology. Academic, New York, pp 268–361

Stebbins GL (1971) Chromosomal evolution in higher plants. Addison-Wesley, Menlo Park, California

Therman E (1972) Chromosome breakage by 1-methyl-2-benzylhydrazine in mouse cancer cells. Cancer Res 32: 1133–1136

III
Mitotic Cycle and Chromosome Reproduction

Significance of Mitosis

In mitosis the genetic material of a cell is divided equally and exactly between two daughter cells. Each chromosome replicates in interphase, then divides into two daughter chromosomes that segregate in anaphase and, with the other chromosomes of the set, form two daughter nuclei. In other words, the cells undergo a regular alternation of chromosome replication and segregation. The main features of mitosis are strikingly similar and may be observed throughout eukaryotes, from unicellular organisms to man. Figure III.1 illustrates mitosis in untreated human lymphocytes. (Nowadays, only a few cytogeneticists have even seen human chromosomes that have not been treated with colchicine and hypotonic solution.) For comparison, Fig. IX.1 shows the last premeiotic mitosis in untreated pollen mother cells of the liliaceous plant, *Eremurus* (*2n* = 14).

Interphase
The mitotic cycle consists of the mitotic stages and the interphase. In interphase the chromosomes are long despiralized threads that are individually indistinguishable. Interphase nuclei come in diverse shapes from nearly spherical or oblong to lobed or branched structures. These last two forms are exceptional but have been seen, for instance, in many insects. Under the light microscope interphase nuclei appear more or less evenly stained with certain condensed chromosome segments, the so-called chromocenters, which stand out. The shape and size of chromocenters vary greatly among not only species but also different tissues of the same organism.

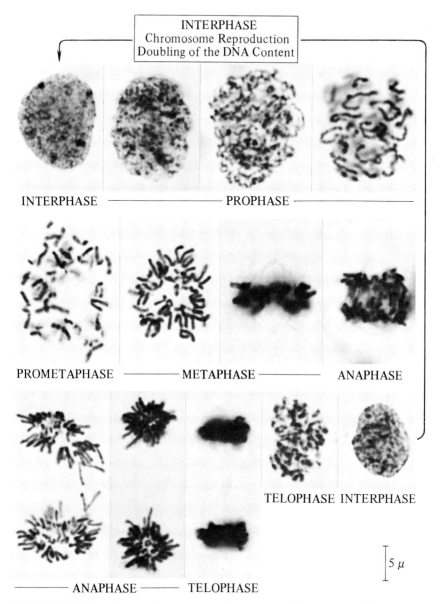

Fig. III.1. Mitotic cycle in cultured human lymphocytes (Feulgen squash).

One or more nucleoli are usually visible in an interphase nucleus. Since the nucleoli tend to fuse, the largest number observed reflects the true number of active nucleolar organizers in the organism.

An interphase nucleus has often been described as the resting nucleus, a singularly inappropriate term, since most of the biochemical activity of the nucleus takes place during this phase. Perhaps interphase nuclei in differentiated tissues "rest" more than the interphases in cells that are still dividing.

Prophase

In prophase, the chromosomes first become visible as thin long threads that gradually shorten and thicken as the diameter of the chromosome coil increases (Figs. III.1 and IX.1). At the same time that the threads are shortening, the nucleoli vanish. If the nucleus is suitably oriented, it is possible to see that the chromosomes reappear in the same positions that they occupied earlier when they despiralized in telophase. The centromeres point in one direction.

Prometaphase

Prophase is followed by a short prometaphase. During this period the nuclear membrane dissolves and the chromosomes, which are nearing their maximum condensation, collect on a metaphase plate.

Metaphase

Outside the nucleus, an organelle called the *centriole* or *centrosome* has divided and the mitotic spindle develops between the centrioles. The centromeres of the chromosomes collect halfway between the poles (centrioles) to form a metaphase plate (Fig. III.1). Long chromosome arms often stick out of the plate. Even though plants do not seem to possess defined centrioles, the spindle arises between two polar areas.

Anaphase

The centromeres divide and the spindle fibers drag the sister chromatids to opposite poles (Figs. III.1 and IX.1).

Telophase

Nuclear membranes are formed around the two chromosome groups. Gradually the chromosome coils loosen, the individual chromosomes become indistinguishable (Figs. III.1 and IX.1), and the nucleoli reappear. Telophase is usually followed by cell division, after which the nuclei revert once more to interphase.

Nondisjunction and Loss of Chromosomes

For various reasons the orderly segregation of daughter chromosomes in anaphase may sometimes fail. Like other biological processes, it is likelier to go wrong in older persons. The inclusion of *both* daughter chromosomes in the *same* nucleus, whatever the mechanism or reason for it may be, is called *nondisjunction*. As a result, one daughter cell gains an extra chromosome, becoming trisomic, whereas the other loses a chromosome and ends up being monosomic.

One daughter chromosome—or sometimes both—may lag behind in its division and not reach either pole. Such laggard chromosomes form micronuclei in interphase. These nuclei usually do not divide, and consequently the chromosome(s) included in them are lost from the complements of both daughter cells.

If the trisomic and monosomic cells arising through nondisjunction or chromosome loss are viable in somatic tissues, the result is mosaicism. Mosaicism may occur in a tissue or, if nondisjunction takes place very early in development, the entire organism may be a mosaic. Actually we are all mosaics to some degree. However, slight mosaicism apparently has no effect on the phenotype. In a female with the cell lines 45,X, 46,XX, and 47,XXX, mosaicism undoubtedly arose through nondisjunction of an X chromosome in the original 46,XX cell line. Loss of a sex chromosome would result in a mosaic for two cell lines 45,X and 46,XX (in a female) or 46,XY (in a male). This form of mosaicism is not too rare.

Mitotic Cycle

The duration of the mitotic cycle varies greatly in different organisms and in different tissues. The cleavage divisions of a toad take 15 minutes, whereas the same process in mouse ear epidermis lasts more than 40 hours (cf. White, 1973). The general rule seems to be that the larger the amount of nuclear DNA, the longer the duration of the mitotic cycle. Thus organisms that are polyploid or have an otherwise high nuclear DNA content display the longest cycles.

The relative lengths of the individual phases also vary, although less than the variability in the duration of the whole cycle. Interphase usually lasts much longer than mitosis. Of the interphasic stages, Gap 1(G_1), which lies between the end of telophase and the beginning of the Synthesis(S) periods, is the longest; DNA synthesis (chromosome replication) takes place during the S period. This stage is succeeded by Gap 2 (G_2), which lasts until prophase. A chromosome is made up of one chromatid during G_1 and two in the G_2 phase.

As already mentioned, the number of haploid chromosome sets in a cell is indicated by the symbols *n, 2n, 3n,* and so on. The symbol for the amount of DNA, as opposed to the number of chromosome sets, is C. The haploid nucleus of a gamete has the DNA content of 1C. A diploid nucleus from anaphase through G_1 has a 2C amount of DNA. This is duplicated during the S period so that from G_2 to metaphase the DNA content of the cell is 4C. It is more practical to talk about 2C and 4C nuclei than about diploid and tetraploid cells, since 4C represents both the G_2 phase of a diploid nucleus and the G_1 stage of a tetraploid one (Patau and Das, 1961).

Chromosome Replication

The mechanics of chromosome reproduction can be studied by autoradiographic techniques, which will be described in more detail in Chapter V. Cells are grown for one S period in medium containing a radioactive constituent of DNA, usually ^{3}H-thymidine, and are then transferred back to normal medium. The results of such an experiment can be seen in Fig. III.2 (cf. Taylor, 1963). Before the S period, the chromosome consists of one double helix of DNA. During the S period, each strand acts as a template for a new radioactive strand. In the subsequent metaphase each chromatid, which now has one "cold" and one "hot" DNA strand, is covered with silver grains. In the following S period, both hot and cold strands act as templates for new cold strands. In metaphase, one chromatid of each chromosome displays silver grains, the other does not. In *diplochromosomes,* which have undergone the same two syntheses without an intervening mitosis and chromatid segregation, the outer chromatids are radioactive (Fig. III.2). This shows that the new chro-

Fig. III.2. Semiconservative replication of eukaryotic chromosomes demonstrated with tritiated thymidine and autoradiography.

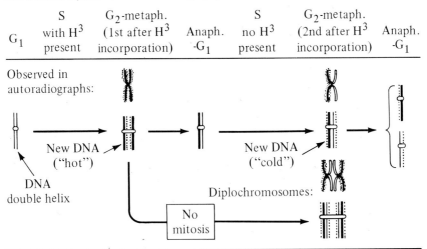

Conclusions: (i) at least one double helix per chromatid
(ii) linear organization at centromere: new DNA at outside

matids are the innermost in a diplochromosome. In the third diploid mitosis after the ^{3}H-thymidine treatment, the hot and the cold chromatids are distributed at random among the chromosomes (Fig. III.2). This type of chromosome reproduction is called semiconservative.

Another characteristic feature in the replication of human and other eukaryotic chromosomes is that the DNA synthesis starts at several points along the chromosome. This can be demonstrated by giving the cells a short pulse of ^{3}H-thymidine after which they are grown in cold medium. In metaphase the chromosome shows labeling over separate short stretches. The various chromosome segments start their replication at different times during the S period, with the heterochromatic segments of chromosomes being the last to replicate. The replication of chromosome segments can be studied more accurately by the more recent 5-bromodeoxyuridine techniques (cf. Latt, 1979).

References

Latt SA (1979) Patterns of late replication in human X chromosomes. In: Vallet HL, Porter IH (eds) Genetic mechanisms of sexual development. Academic, New York, pp 305–329

Patau K, Das NK (1961) The relation of DNA synthesis and mitosis in tobacco pith tissue cultured in vitro. Chromosoma 11: 553–572

Taylor JH (1963) The replication and organization of DNA in chromosomes. In: Taylor JH (ed) Molecular genetics I. Academic, New York, pp 65–111

White MJD (1973) Animal cytology and evolution, 3rd edn. University Press, Cambridge, England

IV
Modifications of Mitosis

Although normal mitosis is characterized by the regular alternation of chromosome reproduction and segregation of daughter chromosomes, the two processes are not necessarily correlated and their relationship can be changed in different ways (cf. Oksala, 1954). Such modifications lead, as a rule, to a chromosome constitution differing from the basic complement of the individual. Apart from multipolar mitoses, all other mitotic modifications are characterized by an absent or defective spindle, and in most cases these result in the duplication of the chromosome number. The terminology referring to mitotic modifications has been the subject of considerable dispute. Present usage is in accord with that proposed by Levan and Hauschka (1953), since it is established in the literature.

Endoreduplication

The most common modification of mitosis is *endoreduplication,* in which the chromosomes replicate two or more times between two mitoses instead of once as in normal mitosis (cf. Levan and Hauschka, 1953). If two replications have occurred, the chromosomes in a subsequent mitosis consist of four chromatids instead of the usual two. As mentioned in an earlier chapter, such structures are called diplochromosomes (Fig. XV.4). After three or four endoreduplications, bundles consisting of 8 or 16 chromatids, respectively, can be observed.

In differentiated cells one endoreduplication after another may take place, so that the nucleus increases stepwise in both size and degree of polyploidy. Enormous nuclei, which have obviously come about through repeated endoreduplications, have been observed. These nuclei are characteristic of tumor cells, in both mammalian cancers (Fig. IV.1) and plant tumors.

Endoreduplication probably occurs occasionally in all tissues. For instance, in human fibroblast cultures 3–5 percent of the dividing cells show a tetraploid chromosome number (a few are even octoploid),

Fig. IV.1.   (a) Half of a giant nucleus from a human cervical cancer cell compared with (b) diploid stroma nuclei. (c) Sideview of a tripolar metaphase from a mouse cancer cell.

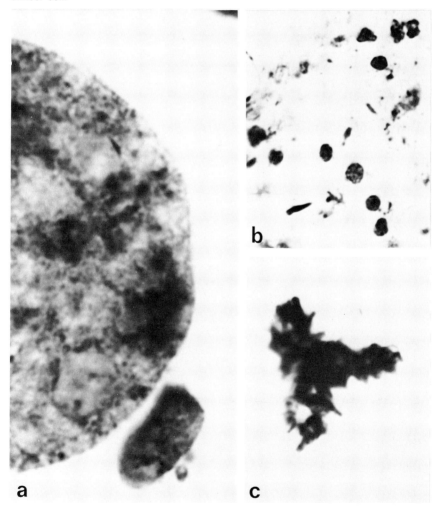

whereas such divisions are rare in cultured lymphocytes. Diplochromosomes are only found in a fraction of the tetraploid divisions, however, since they occur only in the first division after endoreduplication. In many differentiated tissues, on the other hand, endoreduplication is a common phenomenon, as described in the following paragraphs.

Endomitosis

In *endomitosis,* prophase is not abnormal. However, the nuclear membrane never dissolves and the chromosomes continue to contract within it (Fig. IV.2). After they have reached their maximum contraction in

Fig. IV.2. Modifications of mitosis (see text) (Oksala and Therman, 1974).

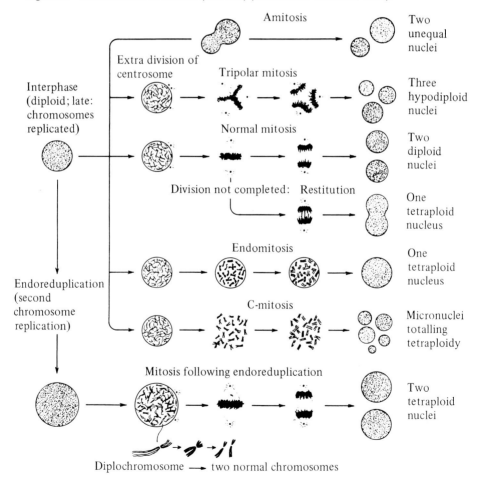

endometaphase, the sister chromatids separate in endoanaphase but do not move far apart. The chromosomes undergo telophasic changes and the nucleus reverts to interphase, having doubled its chromosome number.

Endomitosis was first described in a heteropteran insect, *Gerris,* by Geitler (1939). In this insect the salivary gland cells become 1024- to 2048-ploid as a result of successive endomitoses. Endomitosis is found in the tissue of many other insects as well as other animal groups (cf. Geitler, 1953; Heitz, 1953). The tapetum cells of the anther lobes in many plants undergo endomitosis (cf. D'Amato, 1952; Oksala and Therman, 1977). It has also been described repeatedly in cancer cells of both mouse and man (Levan and Hauschka, 1953; Oksala and Therman, 1974), but the only *normal* mammalian tissue in which it has been so far observed is human placenta (Therman, unpublished).

Polytene Chromosomes

In most genetics and cytogenetics textbooks, polytene chromosomes are represented by the giant chromosomes in dipteran salivary glands. Modifications of these typically giant chromosomes are found in many other insect tissues (cf. Beerman, 1962). A polytene chromosome is composed of uncoiled chromosome strands paired side-by-side. At certain points, called *chromomeres,* the strand is folded. The chromomeres are aligned and form the visible bands in the polytene chromosome. In the diptera, each polytene chromosome consists of the two homologues. The stretched-out parts of the chromosome strands are the interband regions, whereas each band contains one gene. The chromosomes replicate time and again, and a dipteran salivary chromosome may finally consist of as many as 16,000 units (Beerman, 1962).

During the development of a tissue, certain bands specific for the tissue form so-called puffs. Puffing involves the unraveling of the coiled DNA of a particular chromomere; RNA is being synthesized at the puffs, which demonstrates that the puffed genes are active during specific periods in development.

Polytene chromosomes are not limited to insects but are found in certain specialized plant cells, such as the synergids or antipodal cells in the embryo sac (cf. Tschermak-Woess, 1956). Interestingly, typical polytene chromosomes have recently been described in the giant trophoblasts of the rabbit placenta (Zybina et al, 1975). The banded structure is usually much less clear in nondipteran polytene chromosomes, since only in diptera do mitotic chromosomes tend to pair. However, very good banding has been obtained in the bean *Phaseolus* by growing the plants at low temperatures (cf. Nagl, 1976b).

C-Mitosis

In C-mitosis (named after the drug, colchicine) the chromosomes behave normally throughout prophase and into metaphase. However, the spindle is defective or absent so that the chromosomes do not collect into a metaphase plate. Scattered around the cell, they divide and thereafter either unite into one tetraploid restitution nucleus or form a number of micronuclei of variable sizes (Fig. IV.2). Such micronuclei are usually unable to divide further. C-mitosis is extremely rare in normal cells, but it has been described repeatedly in mammalian cancer cells where it probably reflects anoxia and tissue degeneration.

Restitution

Still another mitotic abnormality that leads to the doubling of the chromosome number is restitution. If the two groups of daughter chromosomes fail to separate in anaphase, for example, they may form one dumbbell-shaped nucleus (Fig. IV.2). Restitution may also take place in prophase or metaphase; or, if a cell has two nuclei that divide simultaneously (a common phenomenon in plant tapetum cells), two metaphase or anaphase plates may fuse. Restitution may very occasionally occur in normal human cells, but it is not rare in malignant tissue.

Multipolar Mitoses

A mitotic aberration that can be regarded as the reverse of an absent or defective spindle is the formation of multipolar spindles; the most common are spindles with three poles (Fig. IV.1c). The next in frequency are quadripolar mitoses, and spindles with increasing numbers of poles are correspondingly rarer. Most tripolar and quadripolar spindles come about by one or both centrosomes dividing twice during a mitotic cycle, and not by cell and nuclear fusions as assumed earlier (Therman and Timonen, 1950). Like so many other mitotic abnormalities, multipolar divisions are frequent in malignant tumors but practically nonexistent in normal cells (cf. Oksala and Therman, 1974). If a multipolar anaphase is followed by cell division (often prevented by restitution), the resulting three or more daughter cells usually have abnormal chromosome constitutions and will rarely divide again.

Methods of Studying Endopolyploidy

Most mitotic modifications result in the duplication or multiplication of the chromosome complement. This condition is called *somatic polyploidy* or *endopolyploidy.* Endoreduplication can be inferred from the presence of diplochromosomes or chromatid bundles of even higher multiplicity in a subsequent mitosis. Although most differentiated cells have ceased to divide, endopolyploidy is indicated by the increased size of nuclei, which fall into definite classes.

In many cases, differentiated cells can be induced to undergo mitosis either by wounding or by suitable chemical treatment. For instance, the chromosomes in mouse liver, in whose development endopolyploidy seems to play an important role, can be studied by removing a part of the liver and fixing the cells when the tissue begins to regenerate. Mouse liver cells can also be induced to divide by injecting the animal with carbon tetrachloride (CCl_4).

Differentiated plant cells also renew their mitotic activity around a wound or after treatment with plant-growth substances (cf. Therman, 1951; D'Amato, 1952; Tschermak-Woess, 1956).

Apart from increased size, another indication of polyploidy in differentiated nuclei is the increased size or number of heterochromatic clumps, the chromocenters. The amount of DNA, and thus the degree of polyploidy, can also be determined by measuring spectrophotometrically the dye content of nuclei stained by the Feulgen method or with other stains whose intensities are proportional to DNA content (Patau and Srinivasachar, 1960).

Somatic Polyploidy and Differentiation

Although somatic polyploidy probably occurs occasionally in all meristematic tissues, it is much more common in differentiated tissues. In certain groups of organisms somatic polyploidy obviously plays an important role in differentiation itself. Endopolyploidy has been observed in some differentiated tissues of all major organism groups, both plants and animals, in which it has been looked for (D'Amato, 1952, 1964; Geitler, 1953; Heitz, 1953; Tschermak-Woess, 1971; Brodsky and Uryvaeva, 1977). However, some genera or even whole families, such as Compositae, seem to be devoid of it (cf. Tschermak-Woess, 1971).

Plants

Differentiated plant tissues are usually mosaics of diploid, tetraploid, and more highly polyploid cells. Consider an onion root. In the growing point, which occupies a few millimeters of the tip, the cells are diploid and divide normally. Above this meristematic region is a zone where cell divisions have stopped. The cells grow in size and the nuclei attain various degrees of ploidy through endoreduplication. Above this region the differentiated cells, which do not change further, form a mosaic of diploid, tetraploid, octoploid, and even more highly polyploid cells.

In certain plants, the best studied of which is spinach, cells still divide in the zone of differentiation, and the metaphases exhibit *2n, 4n,* and *8n* diplochromosomes in addition to diploid chromosome constitutions.

Insects

In insects, somatic polyploidy is intimately connected with tissue differentiation, as first pointed out by Oksala (1939) in a study on the tetraploid connective and fat tissues in dragonflies. Such observations on somatic polyploidy have increased to the extent that it seems justified to state that the entire process of differentiation in insects is correlated with the occurrence of endopolyploidy, the various tissues having their own characteristic degrees of ploidy. Some tissues consist of one type of cells, whereas others are mosaics of cells with different ploidy (cf. White, 1973).

Mammals

There are very few observations of somatic polyploidy in differentiated mammalian cells. It was previously mentioned that endoreduplication is obviously a factor in the differentiation of mammalian liver cells. Mammalian megakaryocytes also seem to be highly polyploid. Endopolyploidy is also present in mammalian placental cells. Cancer cells are often characterized by very high degrees of polyploidy (Fig. IV.1).

All in all, the cytological study of mammalian somatic cells is in its infancy and may yet yield results as interesting as the surprising observation of giant polytene chromosomes in the trophoblasts of the rabbit referred to earlier (Zybina et al, 1975).

The Role of Somatic Polyploidy in Differentiation

That there is a correlation between endopolyploidy and cell differentiation is obvious. However, the exact relationship between these two phenomena is unclear. In insects, in whom each tissue may represent a specific level of ploidy, it seems justified to assume that the differentiation depends directly on endopolyploidy. However, in mosaic tissues like the onion root already described, the relationship is much more difficult to interpret. To complicate the issue further, an aberrant onion root may occasionally have a tetraploid basic chromosome constitution instead of a diploid one, but its structure will still be normal. The same is true of corresponding organs in a diploid and a tetraploid variety of a plant species.

Beerman (1962) suggested that it was more economical to amplify the number of genes in dipteran tissues by the formation of polytene chromosomes than by the usual process of chromosome replication, mitosis, and cell division. This idea agrees with Nagl's interesting observation (1976a) that endopolyploidy reaches high levels in organisms, such as dipteran insects and the plant *Capsella*, in which the basic amount of DNA per nucleus is small, but the condition is nonexistent in conifers or amphibians with a large amount of DNA per nucleus. Nagl's conclusion is that somatic polyploidy is a way to compensate for a small basic amount of cellular DNA.

It is a well-known phenomenon that in unicellular organisms like the green alga *Micrasterias,* the duplication of chromosome number is reflected in both morphological and physiological changes in a cell (cf. Kallio, 1953). This phenomenon is probably the result of certain duplicated genes having much more pronounced effects than others.

In dipteran salivary nuclei, heterochromatic chromosome segments reproduce fewer times than the rest of the chromosomes (cf. Beerman, 1962). A similar explanation may apply to the observation that when differentiated cells in the onion root are induced to divide with growth substances, the chromosomes in the polyploid—but not the diploid— nuclei exhibit numerous structural abnormalities, such as fragments, dicentrics, and rings (Therman, 1951). This may be the result of selective duplication of certain chromosome segments which in turn could be a factor in cell differentiation.

The uncertainties surrounding the relationship of endopolyploidy and differentiation can best be summarized in the words of Nagl (1976b, p. 61): "It is also not certain whether endopolyploidy is the result of differentiation, stabilized because of its high selective advantage, or a prerequisite of the mode of differentiation found in angiosperms, insects, and other groups."

References

Beerman W (1962) Riesenchromosomen. In: Protoplasmatologia, Handbuch der Protoplasmaforschung, Vol VI, D. Springer, Vienna

Brodsky WY, Uryvaeva IV (1977) Cell polyploidy: its relation to tissue growth and function. Int Rev Cytol 50: 275–332

D'Amato F (1952) Polyploidy in the differentiation and function of tissues and cells in plants. Caryologia 4: 311–358

D'Amato F (1964) Endopolyploidy as a factor in plant tissue development. Caryologia 17: 41–52

Geitler L (1939) Die Entstehung der polyploiden Somakerne der Heteropteren durch Chromosomenteilung ohne Kernteilung. Chromosoma 1: 1–22

Geitler L (1953) Endomitose und endomitotische Polyploidisierung. In: Protoplasmatologia, Handbuch der Protoplasmaforschung, Vol VI, C. Springer, Vienna

Heitz E (1953) Über intraindividuale Polyploidie. Arch Julius Klaus-Stiftung 28: 260–271

Kallio P (1953) On the morphogenetics of the desmids. Bull Torrey Bot Club 80: 247–263

Levan A, Hauschka ST (1953) Endomitotic reduplication mechanism in ascites tumors of the mouse. J Natl Cancer Inst 14: 1–46

Nagl W (1975a) DNA endoreduplication and polyteny understood as evolutionary strategies. Nature 261: 614–615

Nagl W (1976b) Nuclear organization. Ann Rev Plant Physiol 27: 39–69

Oksala T (1939) Über Tetraploidie der Binde- und Fettgewebe bei den Odonaten. Hereditas 25: 132–144

Oksala T (1954) Timing relationships in mitosis and meiosis. Caryologia (Supp) 6: 272–281

Oksala T, Therman E (1974) Mitotic aberrations in cancer cells. In: German J (ed) Chromosomes and cancer. Wiley, New York, pp 239–263

Oksala T, Therman E (1977) Endomitosis in tapetal cells of Eremurus (Liliaceae). Am J Bot 64: 866–872

Patau K, Srinivasachar D (1960) A microspectrophotometer for measuring the DNA-content of nuclei by the two wave length method. Cytologia 25: 145–151

Therman E (1951) The effect of indole-3-acetic acid on resting plant nuclei. I. *Allium cepa*. Ann Acad Sci Fenn A IV 16: 1–40

Therman E, Timonen S (1950) Multipolar spindles in human cancer cells. Hereditas 36: 393–405

Tschermak-Woess E (1956) Karyologische Pflanzenanatomie. Protoplasma 46: 798–834

Tschermak-Woess E (1971) Endomitose. In: Handbuch der Allgemeinen Pathologie. Springer, Heidelberg pp 569–625

White MJD (1973) Animal cytology and evolution, 3rd edn. University Press, Cambridge, England

Zybina EV, Kudryavtseva MV, Kudryavtsev BN (1975) Polyploidization and endomitosis in giant cells of rabbit trophoblast. Cell Tiss Res 160: 525–537

V
Methods in Human Cytogenetics

In this chapter the usages of the various human chromosome techniques are reviewed. For the actual recipes the reader is referred to Darlington and La Cour (1976), to *Human Chromosome Methodology* (Yunis, 1965, 1974), and to Dutrillaux and Lejeune (1975).

Direct Methods

In the 1920s and earlier, work on human chromosomes was mostly done on testicular tissue, using paraffin-sectioned preparations stained with hematoxylin. One can only admire the fact that with these relatively primitive techniques the human cytologists came as close to the right chromosome number as 48.

In the late 1940s the Feulgen squash technique came into general use (cf. Darlington and La Cour, 1976). It was an excellent method for studying unusual mitoses—for instance, the analysis of mitotic aberrations in cancer cells. The chromosome constitution can also be analyzed in Feulgen squash preparations, if the number of chromosomes is not too high and the metaphase plate not too crowded. The root tips and pollen mother cells of many plants as well as the testicular tissue of animals have been studied successfully using this method.

The squash technique was greatly improved by the added step of treating the cells with drugs before fixation. These drugs, particularly colchicine, shorten the chromosomes, destroy the mitotic spindle, and as a result spread the chromosomes around the cell. Colchicine also prevents cells from entering anaphase, which leads to an accumulation of metaphases, a distinct advantage in cytological work. The most common fixative for Feulgen preparations, as well as for cells in tissue culture, is acetic acid-ethanol.

Tissue-Culture Techniques

The real breakthrough in human cytogenetics came, however, when tissue cultures were combined with colchicine prefixation and treatment with hypotonic solution. Thus Tjio and Levan (1956) were able to report that in cultured embryonic lung cells the human chromosome number was 46. It should be pointed out to the chagrin of everyone who had already been working on human chromosomes—including this author—that colchicine had been used to spread plant chromosomes and to produce polyploidy in plant breeding for two decades and that hypotonic solution swells animal cells had been known since the last century.

Until 1960 when Moorhead et al launched the short-term lymphocyte culture technique, human chromosomes were mainly studied in cultured fibroblasts and in bone marrow cells. The essential agent in lymphocyte cultures is a phytohemagglutinin, of which the most widely used is the extract of kidney beans. Leukemic lymphocytes divide in vitro without "phyto," as it is usually called, whereas normal lymphocytes do not. Short-term lymphocyte cultures provide an important tool in human chromosome studies and they have also found wide use in the cytogenetic studies of diverse animals. It takes only 48–72 h to get chromosome preparations with this easy method, whereas weeks of growth are needed before chromosomes can be studied in fibroblast cultures.

Drying the cells on the microscope slide was a further improvement; it makes the staining and photographing of the flattened chromosomes much more convenient. The most common chromosome stains have been orcein and Giemsa, but azur A and Feulgen as well as a few others have also been used.

At present, cultured lymphocytes and fibroblasts are the main sources of cells for human chromosome studies. However, both types of cultured cells have the disadvantage that they do not reveal what happens in vivo. This can be seen in mitoses from direct bone marrow biopsies that, however, usually do not yield as beautiful chromosome preparations as cultured cells.

Amniocentesis, which has become an important tool in genetic counseling, involves taking a sample of amniotic fluid from a pregnant woman and preparing a tissue culture from cells centrifuged from it. The fetal chromosomes can be analyzed from such a culture and an abortion performed if the cells reveal a chromosome anomaly that would lead to serious phenotypic consequences. Certain metabolic diseases can also be determined by biochemical tests on fetal cells.

Meiotic Studies

Meiosis has been studied far more extensively in the human male than in the female. Initially male meiosis was usually analyzed from biopsies made into Feulgen squash preparations. A more recent method is to prepare the meiocytes out of the testicular tubules and swell them in hypotonic solution, after which they are fixed and dried on a slide in the same way as lymphocytes.

Female mammalian meiosis has been analyzed, especially in the mouse. Similar attempts have been made in man (Edwards, 1970; Jagiello, et al, 1976), but so far the results have not been comparable with those obtained in the mouse. Meiotic chromosome preparations of the mouse and other animals can be improved by an injection of colchicine into the animal a few hours before it is sacrificed. Meiotic prophase stages have been studied successfully in human embryonic oocytes (cf. Therman and Sarto, 1977). A special technique for the dictyotene (diplotene) stage in man has been developed by Stahl et al (1976).

Sex Chromatin Techniques

Since all the human cell types that are generally cultured are of mesodermal origin, practically the only information on the ectoderm and endoderm comes from studies on sex chromatin. This includes the X chromatin (Barr body), which consists of the inactive X, and the Y chromatin body, which represents the distal end of Yq. These are most often determined in cell smears made from the buccal mucosa. Barr bodies are usually stained with orcein, acid fuchsin, or Feulgen (Fig. XII.2), but the fluorochrome acridine orange also gives excellent results (cf. Fig. 2 in Dutrillaux, 1977). Because the Y body is too small to be distinct under the light microscope, it is usually stained with quinacrine and studied by fluorescence microscopy (Fig. XII.1). The fluorescent Y body is also visible in Y-carrying sperm. In addition to buccal smears, X chromatin has been analyzed in cultured fibroblasts, hair-root cells, and cells of the vaginal epithelium.

Autoradiography

As already discussed, only the chromosomes 1, 2, 3, 16, 17, 18, and the Y can be individually distinguished under the light microscope, the rest being identifiable only with respect to their groups. The realization that chromosomes and chromosome segments replicate at different times during the S period contributed a considerable step forward in chromosome identification.

For autoradiography, the cells are fed a radioactive nucleotide (usually ^{3}H-thymidine) 5–9 h before fixation. This means that most cells observed in metaphase were then in the latter part of the S period. Those chromosome segments still replicating take up the radioactive thymidine. The cells are fixed and treated as usual and a number of suitable metaphases photographed. The slides are dipped in photographic emulsion, exposed in the dark, and developed. The previously photographed cells are rephotographed if they show a suitable number of silver grains (Fig. XII.3a and b), which reflect the uptake of the radioactive compound. This method has been important in distinguishing the inactive X chromosome(s) from the active one. Some other chromosomes, for instance 21 and 22, can also be separated because 21 replicates later.

However, autoradiography, which is both time consuming and inaccurate, has been practically replaced by banding techniques (for chromosome identification) and by the recent bromodeoxyuridine (BrdU) techniques (for the analysis of the order of chromosome replication). Autoradiography is still used in cytogenetics to determine the timing of DNA synthesis as well as of the duration of the mitotic stages.

Autoradiography plays an important role in the in-situ hybridization techniques. They are used in the localization of specific, clustered, repeated DNA sequences on the chromosomes. This is done by hybridizing either ^{3}H-DNA labeled in vivo or ^{3}H-RNA transcribed in vitro with bacterial RNA polymerase directly on the chromosome slides (Pardue and Gall, 1970). Thereafter autoradiography is performed in the usual way and the location of the silver grains indicates the site of the specific DNA (Pardue, 1975).

Banding Techniques on Fixed Chromosomes

A decisive step forward in human cytogenetics was the invention of banding techniques that differentiate the chromosomes into transverse bands of different lengths. With these methods all the chromosomes of man and many other organisms, and even the breakpoints in most structural rearrangements, can be identified. The banding techniques and what they reveal about chromosome structure have been reviewed in many articles, for example Arrighi (1974), Latt (1976), Evans (1977), Sanchez and Yunis (1977), Dutrillaux (1977), and Ris and Korenberg (1979).

When human chromosomes are stained with quinacrine HC1 (Atabrine) or quinacrine mustard and studied with a fluorescence microscope, they show bands of different brightness (Caspersson et al 1970). These are called *Q-bands* (quinacrine bands) (Figs. XIV.1 and 3). Slides stained with quinacrine are not permanent and after a couple of photomicrographic exposures, the fluorescence fades enough to be unusable.

The other type of banding techniques involves various pretreatments and staining with Giemsa or certain fluorochromes; Giemsa bands *(G-bands)* are obtained when the chromosomes are pretreated with a salt solution at 60°C or with proteolytic enzymes, usually trypsin. Giemsa banding yields essentially the same information as Q-banding, only the brightly fluorescent Q-bands are now darkly stained, whereas the Q-dark regions are now light (Fig. II.4). Each method has its advantages. With Q-banding the chromosomes are stained without any pretreatment and their morphology is retained. This makes measurements of bands more accurate. Also the relative brightness of the bands can be estimated (Kuhn, 1976). However, the chromosomes can only be analyzed from photographs.

The G-banded slides, on the other hand, are permanent and are therefore more suitable for routine work. By means of these two techniques some 300 bands have been described in the human chromosomes (Figs. V.1 and XIV.3) (Paris Conference, 1971). However, in most metaphase plates only a fraction of this number of bands can actually be seen. By accumulation of cells in prophase and by banding of the long chromosomes, the number of visible bands has been increased to about 1000 (cf. Sanchez and Yunis, 1977).

Reverse banding *(R-banding)* involves pretreatment with hot (80–90°C) alkali and subsequent staining with Giemsa or a fluorochrome. As the name indicates, the banding pattern is the reverse of G-banding; in other words the bands that are dark with R-banding are light with G-banding and vice versa (Fig. VI.1f). Fluorescent R-banding again is the reverse of Q-banding. Although this banding reveals nothing new, compared with Q-banding and G-banding, it complements them when chromosome ends are espcecially studied, as in distal deletions and translocations. A modification of R-banding, called T-banding, brings out mainly the tips of chromosomes.

Another technique, which also utilizes the Giemsa stain, has given additional information about chromosome structure. Centric banding *(C-banding)*, for which chromosomes are usually first treated in acid and then in alkali (barium hydroxide, for example) prior to Giemsa staining, brings out the heterochromatic regions around the centromere (Fig. VI.1f and g) and in the distal end of the Y chromosome. With a modification of the C-banding technique (G-11), it is possible to stain the centric heterochromatin in human chromosome 9 specifically.

The banding technique used most successfully in plants corresponds to C-banding, and the resulting darkly stained bands probably also represent constitutive heterochromatin (Chapter VI).

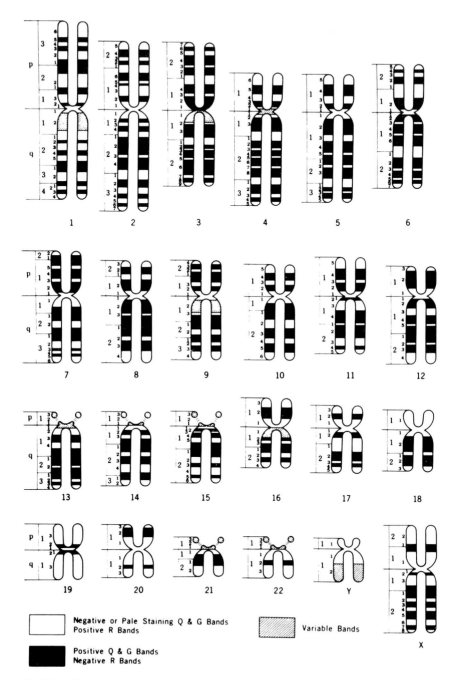

Fig. V.1. Diagram of a banded human karyotype (Paris Conference, 1971).

Negative or Pale Staining Q & G Bands
Positive R Bands

Variable Bands

Positive Q & G Bands
Negative R Bands

New banding techniques, often combining two or more stains, appear frequently. A variety of fluorochromes, including Hoechst 33258, a number of antibiotics, and so-called DAPI and DIPI stains, have found use in the detailed analysis of chromosome bands (cf. Schnedl, 1978).

With a special Giemsa method it is possible to stain nucleolar organizers differentially, and a silver (NOR) staining technique reveals the organizers that are active (Fig. VI.1c and e) (cf. Bloom and Goodpasture, 1976).

Banding Techniques for Cells in Culture

The banding techniques that involve the treatment of cells before fixation are based on the important discovery that chromosomes and chromatids that have incorporated BrdU have a different structure and consequently different staining properties than those containing thymidine. One of the applications of this observation is to cause one chromatid to incorporate BrdU, whereas the other contains thymidine. Sister chromatid exchanges, the existence of which was already revealed by autoradiography, can be studied with much greater accuracy with this technique (Fig. V.2) (Latt, 1973). For the analysis of sister chromatid exchanges, the fluorochromes Hoechst 33258, acridine orange, and coriphosphine O, as well as the Giemsa stain after heat treatment, have been used (Korenberg and Freedlender, 1974).

Substitution with BrdU has also been successful in demonstrating the late-labeling X chromosome and other individual chromosomes and segments that replicate in the latter part of the S period (Latt, 1973, 1974). This can be done, for instance, by growing lymphocytes for 40–44 h in a medium containing BrdU and, thereafter, feeding them thymidine for 6–7 h. When stained with a suitable fluorochrome, the late-replicating X chromosome fluoresces more brightly than the other chromosomes or is darker than the others when Giemsa staining is used. An example of this technique is seen in Fig. XII.4 in which the late-replicating X chromosomes stand out.

Nomenclature of Human Chromosomes

As a continuation to previous agreements on the nomenclature of human chromosomes, in the Paris Conference (1971) a system to identify human chromosome bands and to indicate various chromosome abnormalities was proposed. Figure V.1 shows a diagram of a banded human karyotype according to this system. Telomeres, centromeres, and a number of prominent bands are used as "landmarks." A section of a chromosome

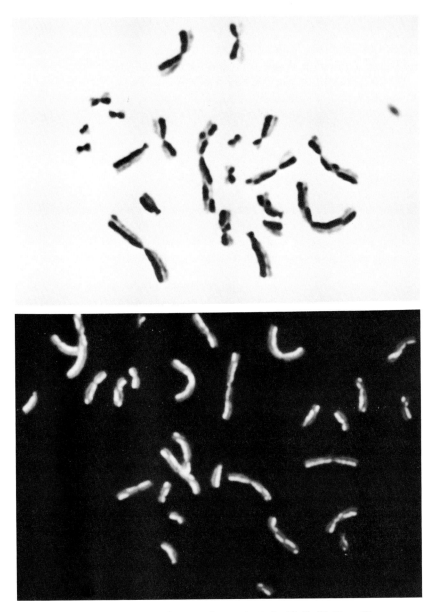

Fig. V.2. Sister chromatid exchanges demonstrated with BrdU. Top: Giemsa-stained Chinese hamster cell (courtesy of JR Korenberg). Bottom: human lymphocyte stained with coriphosphine O.

between two landmarks is called a region and they are numbered 1, 2, 3, and so on in both directions, starting with the centromere. The bands within the regions are numbered according to the same rule. Thus the first band in the second region of the short arm of chromosome 1 is 1p21.

For the designation of chromosome abnormalities, two systems—one short and one detailed—were put forward. For the actual use of both systems the reader is referred to the Paris Conference (1971) and to Sanchez and Yunis (1977). In the following discussion only a few basic examples of the short system are given.

An extra or a missing chromosome is denoted with a plus or a minus sign, respectively, before the number of the chromosome. Thus the chromosome constitution of a female with trisomy for chromosome 13 would be 47,XX,+13, and a male with monosomy for 21 would have the formula 45,XY,−21.

A plus or minus sign after the symbol of the chromosome arm means that a segment is added to or missing from it. For example, a female with the cri du chat syndrome would have the chromosome formula 46,XX,5p− and a male with an abnormally long 4q would be designated as 46,XY,4q+. The karyotype of a female with a Robertsonian translocation (centric fusion) between chromosomes 13 and 14 would be 45,XX,t(13q14q). The formula for a male carrier in whom chromosome arms 3p and 6q have exchanged segments, the breakpoints being 3p12 and 6q34, would be 46,XY,t (3;6) (p12;q34).

Unfortunately the lengths of chromosome bands in Fig. V.1 (Paris Conference, 1971) are not based on actual measurements, although a realistic banded diagram would be most useful. Furthermore, only a small fraction of metaphase plates reveals all the bands illustrated in this diagram. Therefore it is often difficult, if not impossible, to determine the exact breakpoints in, say, a reciprocal translocation. Despite this, most authors feel obliged to specify the breakpoints, even when they are based only on guesses. It may be possible to determine breakpoints more accurately in the future, when the banding of prophase chromosomes comes into wider use (cf. Sanchez and Yunis, 1977).

References

Arrighi FE (1974) Mammalian chromosomes. In: Busch H (ed) The cell nucleus. Academic, New York, pp 1–32

Bloom SE, Goodpasture C (1976) An improved technique for selective silver staining of nucleolar organizer regions in human chromosomes. Hum Genet 34: 199–206

Caspersson T, Zech L, Johansson C (1970) Differential binding of alkylating fluorochromes in human chromosomes. Exp Cell Res 60: 315–319

Darlington CD, La Cour LF (1976) The handling of chromosomes, 6th edn. Wiley, New York

Dutrillaux B (1977) New chromosome techniques. In: Yunis JJ (ed) Molecular structure of human chromosomes. Academic, New York, pp 233–265

Dutrillaux B, Lejeune J (1975) New techniques in the study of human chromosomes: Methods and applications. In: Harris H, Hirschhorn K (eds) Advances in human genetics, Vol 5. Plenum, New York, pp 119–156

Edwards RG (1970) Observations on meiosis in normal males and females. In: Jacobs PA, Price WH, Law P (eds) Human population cytogenetics. Williams and Wilkins, Baltimore, pp 10–21

Evans HJ (1977) Some facts and fancies relating to chromosome structure in man. In: Harris H, Hirschhorn K (eds) Advances in human genetics, Vol 8. Plenum, New York, pp 347–438

Jagiello G, Ducayen M, Fang J-S, et al (1976) Cytogenetic observations in mammalian oocytes. In: Pearson PL, Lewis KR (eds) Chromosomes today, Vol 5. Wiley, New York pp 43–63

Korenberg JR, Freedlender EF (1974) Giemsa technique for the detection of sister chromatid exchanges. Chromosoma 48: 355–360

Kuhn EM (1976) Localization by Q-banding of mitotic chiasmata in cases of Bloom's syndrome. Chromosoma 57: 1–11

Latt SA (1973) Microfluorometric detection of deoxyribonucleic acid replication in human metaphase chromosomes. Proc Natl Acad Sci USA 70: 3395–3399

Latt SA (1974) Microfluorometric analysis of DNA replication in human X chromosomes. Exp Cell Res 86: 412–415

Latt SA (1976) Optical studies of metaphase chromosome organization. Ann Rev Biophys Bioeng 5: 1–37

Moorhead PS, Nowell PC, Mellman WJ, et al (1960) Chromosome preparations of leukocytes cultured from human peripheral blood. Exp Cell Res 20: 613–616

Pardue ML (1975) Repeated DNA sequences in the chromosomes of higher organisms. Genetics 79: 159–170

Pardue ML, Gall JG (1970) Chromosomal localization of mouse satellite DNA. Science 168: 1356–1358

Paris Conference (1971) Standardization in human cytogenetics. Birth Defects: original article Series, VIII: 7. The National Foundation, New York, 1972

Ris H, Korenberg JR (1979) Chromosome structure and levels of chromosome organization. In: Goldstein L, Prescott DM (eds) Cell biology, Vol 2. Academic, New York pp 268–361

Sanchez O, Yunis JJ (1977) New chromosome techniques and their medical applications. In: Yunis JJ (ed) New chromosomal syndromes. Academic, New York, pp 1–54

Schnedl W (1978) Structure and variability of human chromosomes analyzed by recent techniques. Hum Genet 41: 1–9

Stahl A, Luciani JM, Gagné R, et al (1976) Heterochromatin, micronucleoli, and RNA containing body in the pachytene and diplotene stages of the human oocyte. In: Pearson PL, Lewis KR (eds) Chromosomes today, Vol 5. Wiley, New York, pp 65–73

Therman E, Sarto GE (1977) Premeiotic and early meiotic stages in the pollen mother cells of *Eremurus* and in human embryonic oocytes. Hum Genet 35: 137–151

Tjio JH, Levan A (1956) The chromosome number of man. Hereditas 42: 1–6

Yunis JJ (ed) (1965, 1974) Human chromosome methodology, 1st and 2nd edns. Academic, New York

VI
Longitudinal Differentiation of Eukaryotic Chromosomes

Prebanding Studies

Apart from primary and secondary constrictions and satellites, chromosomes of most higher organisms display differential segments, sometimes whole chromosomes, that exhibit a condensation cycle which deviates from the main part of the chromosomes. Heitz, who made a number of basic studies in this field during the late 1920s, named these chromosome parts *heterochromatic* in contrast to the majority of the chromosomes, which he termed *euchromatic* (cf. Passarge, 1979). Chromosomes that are more condensed than the rest have also been called *positively heteropycnotic,* whereas those that are less condensed are *negatively heteropycnotic*. The same chromosome may at some stages be positively, at others negatively, heteropycnotic. The X chromosome in the spermatocytes of grasshoppers is an example of this.

Being out-of-step with the rest of the chromosomes, which is one of the characteristics of heterochromatin, is called *allocycly*. In interphase, heterochromatic chromosome segments often appear as more condensed chromocenters or prochromosomes. As a result of the studies of Heitz, it is known that heterochromatin is genetically inert. The prebanding knowledge of heterochromatin has been reviewed by Vanderlyn (1949).

The first successful experiments designed to make heterochromatic segments visible in metaphase chromosomes were completed in the late 1930s when Darlington and La Cour (1940) discovered chromosome regions in the liliaceous plants *Paris* and *Trillium* which, after cold treatment, appeared thinner and less stained than the rest of the chromosomes. In interphase nuclei the same segments were visible as positively heteropycnotic chromocenters. This phenomenon has been observed in several plants and a few animal species. Darlington and La Cour (1940) pointed out that these differential segments had to be genetically inert, since they varied from plant to plant, and many individuals were heterozygous for them.

A forerunner of the present banding techniques was the acid treatment combined with orcein staining of Yamasaki (1956), which produced excellent banding in orchid chromosomes. However, Yamasaki's interesting work appeared before the time was ripe for it and it did not attract the attention it deserved.

Banding Studies on Human Chromosomes

Q-banding differentiates the human chromosomes into bands of differing length and relative brightness. These two parameters were determined by Kuhn (1976). An interesting property of the metaphasic Q-bands is that they seem to act as units; for instance, each band seems to replicate within a particular limited time during the S period (cf. Ris and Korenberg, 1979). These bands have also shown amazing constancy during evolution. Because the overwhelming majority of the bands are identical in both man and the great apes, one can conclude that they have remained unchanged for at least 20 million years, some of them for considerably longer.

The most brightly fluorescent bands in the human chromosome complement are the distal end of the Y chromosome and the narrow variable bands at the centromeres of chromosomes 3, 4, and the acrocentrics. The human populations are polymorphic for these bright bands, just as they are for the size and fluorescent properties of the satellites.

G-banding and its reverse, R-banding, give essentially the same information as Q-banding. However, R-banding is often useful for the study of structural changes involving chromosome ends that might go undetected with Q-banding or G-banding.

C-banding reveals a type of chromatin that is in principle different from euchromatin. This technique specifically stains constitutive heterochromatin, which is situated at the centric regions of all human chromosomes and at the distal end of the Y chromosome. Particularly prominent blocks of this heterochromatin are found in chromosomes 1, 9, and 16, and these regions show a considerable polymorphism in the population. Possibly a variation in the centric heterochromatin of these chromosomes is more noticeable than variation in the shorter C-bands; however, variation in some of them has also been reported. The constitutive heterochromatin in chromosomes 1, 9, 16, and the Y varies from shorter than average to some three times its usual length.

Constitutive heterochromatin seems especially prone to breakage and rearrangements. However, for the most part the polymorphism of C-bands is probably the result of unequal pairing and crossing over. Nevertheless these changes cannot take place very often, since the C-band variants as a rule are constant from one generation to the next and show normal mendelian inheritance.

In-situ hybridization techniques show that the satellites of the human acrocentric chromosomes contain the nucleolar organizers (Henderson et al, 1972). This has been confirmed by the silver-staining technique (Fig. VI.1c and e) (Goodpasture and Bloom, 1975), which reveals that the nucleolar organizers are situated in the satellite stalks.

Fig. VI.1. (a) Satellite association of human acrocentric chromosomes; (b) centric association of mouse cancer chromosomes; (c) silverstaining of active nucleolar organizers in human D and G chromosomes (courtesy of C Trunca); (d) hetero-chromatic association of mouse diplotene bivalents; (e) satellite association of silver-stained human acrocentrics (courtesy of TM Schroeder); (f) human chromosome 1 (from the left): prometaphase banding, Q-banding, G-banding, R-banding, C-banding (from a heterozygote for the C-band); (g) C-banded human dicentric chromosomes with one centromere inactivated.

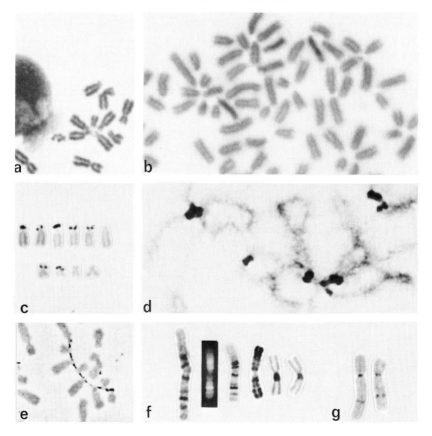

Constitutive Heterochromatin

The characteristics of human heterochromatin are summarized in Table
VI.1. Heterochromatin is divided into two categories, constitutive and
facultative (Brown, 1966). Although these two classes behave similarly
in many ways, a fundamental structural difference exists between them.
The DNA structure that is characteristic of the constitutive heterochromatin is different from that of euchromatic DNA. On the other hand,
facultative heterochromatin consists of essentially euchromatic DNA,
but it *behaves* differently during certain phases of the development.

Constitutive heterochromatin, which in man is located in the C-bands
and at the distal end of the Y chromosome, consists of simple sequence-
repeated DNA that corresponds to the satellite DNAs (cf. Hsu, 1975).
Constitutive heterochromatin contains no mendelian genes and is never
transcribed. This explains the fact that considerable variation of C-bands
does not seem to affect the phenotype, even in extreme cases. Characteristics of constitutive heterochromatin, which it shares with facultative
heterochromatin, are its genetic inertness, its late replication during the
S period and its general allocycly. It has long been known (cf. Vanderlyn,
1949) that heterochromatic segments are "sticky" and tend to fuse in
interphase. This is reflected in the nose-to-nose association of human
acrocentrics (which may be helped by remnants of nucleolar material)
and of mouse chromosomes, both in mitotic metaphase and in diplotene
(Fig. VI.1a,b,d, and e).

Constitutive heterochromatin is found in practically all higher organisms, both plants and animals. It is often situated at the centromere, as
in man and the mouse. In many plants, for instance the onion, it forms
blocks at the telomeres, but it can also be intercalary (Fig. VI.2a). Before
the invention of banding techniques, the deer mouse *Peromyscus* presented a cytological riddle. The length of its chromosomes and the
centromere index seemed to vary from animal to animal, and complicated
inversion systems were invented to explain this observation. C-banding
reveals that all the short arms consist of constitutive heterochromatin,
whose amount varies from animal to animal (cf. Hsu, 1975). Closely
related species with the same chromosome number sometimes have very
different amounts of DNA per nucleus. This, too, has been found to
depend on variation in constitutive heterochromatin.

The reaction of the different C-bands to the various banding techniques
shows that the structure of their DNA is not identical. Thus, although all
the constitutive heterochromatin stains dark with C-banding, the distal
end of the Y chromosome is brightly fluorescent with Q-banding, whereas
the centric heterochromatin of chromosomes 1, 9, and 16 is dark.
Further, a specific staining technique (G-11) has been developed for the
C-band in chromosome 9. This band seems to be more liable to structural
aberrations, especially inversions, than the others.

Table VI.1. Heterochromatin in Man

Class of Heterochromatin	Occurrence	Appearance with			Criteria for Heterochromatin		Late-Replication
		Q-banding	G-banding	C-banding	Heteropycnosis	Genetically inert	
Facultative	Inactive X chromosome(s)	Not affected by inactivation			Barr-body formation, allocycly	Lyon inactivation	+
Constitutive Y-type	Distal Yq	Extremely bright	Variable	+	Positive in interphase, negative in metaphase	Scarcity of Y-linked genes, polymorphism	+
	Variant centric bands and satellites in 3, 4, and acrocentrics		−			Polymorphism	
Centric	Secondary constriction in 1q and 16q	Dark	+	+	Sometimes negative in metaphase	Polymorphism	+
	Secondary constriction in 9q	Dark	Variable	+	Sometimes negative in metaphase	Polymorphism	+
	Centric regions in all chromosomes	Dark	Variable	+		Some polymorphism	
	Short arms and satellites of acrocentrics	Polymorphism for Q-brightness		+	Negative in metaphase	Polymorphism	
Intercalary	In all Q-bright bands	Bright	+	−	Negative in prematurely condensed chromosomes	Imbalance for these segments better tolerated than for others	+

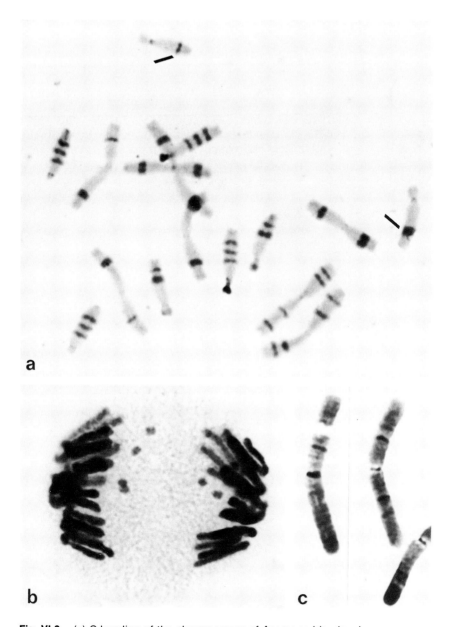

Fig. VI.2. (a) C-banding of the chromosomes of *Anemone blanda;* chromosome with heterozygous C-band marked (courtesy of D Schweizer); (b) B-chromosomes segregating irregularly in I meiotic anaphase of *Fritillaria imperialis* (courtesy of M Ulber); (c) G-banded chromosomes of *Pinus resinosa;* the chromosome on the left has a secondary constriction (courtesy of A Drewry).

That constitutive heterochromatin is a practically ubiquitous constituent of eukaryotic chromosomes must mean that it has an important function. However, the only established effect of heterochromatin is that it seems to regulate crossing-over (cf. Miklos and John, 1979); otherwise we are completely in the dark about its role.

In several plants and animals, but not in man, extra chromosomes, the so-called B chromosomes, have been found (Fig. VI.2b). They occur in addition to the normal chromosome complement and may vary greatly in number from individual to individual within the same species. They appear to be practically inert, since they do not seem to affect the phenotype even when present in considerable numbers (cf. Rees and Jones, 1977). However, they have some effects, especially on viability and chiasma frequency. As a rule, B chromosomes do not show the typical staining of constitutive heterochromatin but appear variably banded, resembling small, normal chromosomes.

Facultative Heterochromatin

One of the best known examples of facultative heterochromatin is the inactive X in the mammalian female. In the human female, one X in each cell becomes inactivated at random during the blastocyst stage. The X behaves thereafter as if it consisted of heterochromatin. It is condensed in interphase and shows no transcription. It replicates late during the S period (Figs. XII.3 and 4) and is more condensed than the other chromosomes, even in prophase and prometaphase (Fig. XII.2). Before inactivation both X chromosomes are active, and the inactive X is turned back on again before oogenesis, when both X chromosomes behave much like the autosomes. During spermatogenesis, on the other hand, neither the X nor the Y chromosome is transcribed. Both chromosomes behave like heterochromatin. The mechanism that determines the behavior of the facultative heterochromatin is still completely unknown.

Intercalary Heterochromatin

A third type of heterochromatin has been called intercalary heterochromatin (Patau, 1973). However, we know even less about it than about the other categories of heterochromatin. It is situated in the Q-bright chromosome bands, which are late replicating and seem to contain fewer genes than the dark bands.

Only a small part of the DNA present is transcribed in any cell (cf. Ris and Korenberg, 1979). No condensed parts, whether consisting of constitutive or facultative heterochromatin, code for any proteins. In some cell types, such as erythrocytes or sperm, in which the total chromatin is in a condensed state (cf. Ris and Kornberg, 1979), this inactivity applies to the whole chromosome complement. A different type of heterochromatic behavior is exhibited by chromosomes or chromosome segments that are condensed in differentiated cells, forming chromocenters of various shapes and sizes. This is obviously a mechanism to shut off those genes whose activity is not needed at a certain point of development (cf. Ris and Korenberg, 1979). However, not only is *condensed* chromatin silent, but a great part of the *extended* chromatin is never transcribed in any cell type either (Ris and Korenberg, 1979).

Chromosome Bands

As new banding techniques continually emerge, often combining different stains, a more and more detailed longitudinal differentiation of chromosomes becomes possible (cf. Schnedl, 1978). Vosa (1976) divided the heterochromatic bands in the long M chromosome of *Vicia* into seven classes according to their reactions to different staining techniques.

However, the ongoing classification of chromosome bands has told us little about their possible functions or about the chemical basis of the differential staining reactions. The most information exists about Q-banding. Weisblum and de Haseth (1972) showed in their in-vitro studies on polynucleotides of known composition that DNA consisting of repeated adenine-thymine (AT) base pairs fluoresces brightly with quinacrine, whereas guanine-cytosine(GC) pairs quench fluorescence. Quenching is also a function of the degree of the interspersion of GC pairs (AT pairs must occur in uninterrupted stretches of a certain length to cause fluorescence). Two stretches of DNA can have the same base ratio (AT:GC) but fluoresce differently, depending on the sequence of the bases.

Korenberg and Engels (1978), who were the first to determine directly the base ratios of the Q-bands in human chromosomes, demonstrated that the Q-brightness of the bands was positively correlated with the AT:GC ratio. However, the differences in the base ratios between the very bright Y heterochromatin and the dark bands are much less than has often been assumed (cf. Evans, 1977). Interestingly, the longer chromosomes are relatively brighter than the smaller ones. Although the DNA structure is the basis for Q-banding, the chromosomal proteins, both histones and nonhistones, obviously play an important role, especially in the various Giemsa techniques (cf. Ris and Korenberg, 1979). Nevertheless our understanding of the chemical basis of the various banding techniques is only just beginning.

Function of Human Chromosome Bands

The in-situ hybridization techniques that were the forerunners of the various G-banding techniques revealed that the eukaryotic chromosomes contain three types of clustered repeated DNA sequences. One type codes for the structural RNAs of the ribosomes; in other words, they are the nucleolar organizers. It is characteristic of these genes to be repeated a few hundred times. An interesting feature is that they seem to have remained unchanged through evolution from *Drosophila* to man. Another type of repeated genes codes for histones, which are important protein components of eukaryotic chromosomes. The third kind of repeated sequences form the constitutive heterochromatin. They should not be called genes, since they never are transcribed (cf. Pardue, 1975).

Chromosome bands that react similarly to the various staining techniques may still reveal different functions and be distinguished on that basis. As a rule, chromosome-breaking agents affect the Q-dark bands, but in a highly nonrandom fashion. Certain regions often stand out as hot spots, whose locations depend on the agent used.

The lymphocytes of patients with Bloom's syndrome exhibit characteristic mitotic chiasmata. Kuhn (1976) demonstrated the extreme nonrandomness of the exchange points in them. Most of the chiasmata were situated in the Q-dark regions, and the short dark regions in 3p, 6p, 11q, 12q, 17q, and 19 (p or q) were outstanding hot spots.

Yunis (1965) pointed out that the only three autosomes (13, 18, and 21) for which trisomy is viable to any appreciable extent are the latest replicating chromosomes in their groups. Now we know that they are also the Q-brightest. Yunis concluded that they contained more heterochromatin than did early replicating chromosomes and, therefore, their trisomies were better tolerated than trisomies for more gene-rich autosomes. A similar nonrandomness has recently been shown for spontaneous trisomic abortions (Table XIV.2) (Boué et al, 1976).

Korenberg et al (1978) found that a strong negative correlation exists between the hotspot chromosomes and their involvement in trisomic abortions. In other words, embryos that are trisomic for these chromosomes die too early to be recognized as spontaneous abortions. Korenberg et al formulated the hypothesis that the Q-darker human chromosome bands have higher gene densities than do the brighter ones, and the hot spots are especially prominent in this respect. Because they contain active genes, they would be looped out in interphase and as a result be more easily available for mitotic pairing and crossing-over. Several individual observations also speak for the relative gene-richness of the Q-dark regions. For instance, trisomy for the short Q-dark distal region of chromosome 21 is responsible for all the symptoms characteristic of Down's syndrome, whereas the bright proximal band gives rise only to mild mental retardation in the trisomic, and even in the mono-somic, state (for example, Hagemeijer and Smit, 1977).

References

Boué J, Daketsé M-J, Deluchat G, et al (1976) Identification par les bandes Q et G des anomalies chromosomiques dans les avortements spontanés. Ann Genet 19: 233–239

Brown SW (1966) Heterochromatin. Science 151: 417–425

Darlington CD, La Cour L (1940) Nucleic acid starvation of chromosomes in *Trillium*. J Genet 40: 185–213

Evans HJ (1977) Some facts and fancies relating to chromosome structure in man. In: Harris H, Hirschhorn K (eds) Advances in human genetics, Vol 8. Plenum, New York, pp 347–438

Goodpasture C, Bloom SE (1975) Visualization of nucleolar organizer regions in mammalian chromosomes using silver staining. Chromosoma 53: 37–50

Hagemeijer A, Smit EME (1977) Partial trisomy 21. Further evidence that trisomy of band 21q22 is essential for Down's phenotype. Hum Genet 38: 15–23

Henderson AS, Warburton D, Atwood KC (1972) Location of ribosomal DNA in the human chromosome complement. Proc Natl Acad Sci USA 69: 3394–3398

Hsu TC (1975) A possible function of constitutive heterochromatin: the body-guard hypothesis. Genetics 79: 137–150

Korenberg JR, Engels WR (1978) Base ratio, DNA content, and quinacrine-brightness of human chromosomes. Proc Natl Acad Sci USA 75: 3382–3386

Korenberg JR, Therman E, Denniston C (1978) Hot spots and functional organization of human chromosomes. Hum Genet 43: 13–22

Kuhn EM (1976) Localization by Q-banding of mitotic chiasmata in cases of Bloom's syndrome. Chromosoma 57: 1–11

Miklos GLG, John B (1979) Heterochromatin and satellite DNA in man: prop-erties and prospects. Am J Hum Genet 31: 264–280

Pardue ML (1975) Repeated DNA sequences in the chromosomes of higher organisms. Genetics 79: 159–170

Passarge E (1979) Emil Heitz and the concept of heterochromatin: longitudinal chromosome differentiation was recognized fifty years ago. Am J Hum Genet 31: 106–115

Patau K (1973) Three main classes of constitutive heterochromatin in man: intercalary, Y-type and centric. In: Wahrman J, Lewis KR (eds) Chromo-somes today, Vol 4. Wiley, New York, p 430.

Rees H, Jones RN (1977) Chromosome genetics. University Park Press, Baltimore

Ris H, Korenberg JR (1979) Chromosome structure and levels of chromosome organization. In: Goldstein L, Prescott DM (eds) Cell biology. Academic, New York, pp 268–361

Schnedl W (1978) Structure and variability of human chromosomes analyzed by recent techniques. Hum Genet 41: 1–9

Vanderlyn L (1949) The heterochromatin problem in cyto-genetics as related to other branches of investigation. Bot Rev 15: 507–582

Vosa CG (1976) Heterochromatin classification in *Vicia faba* and *Scilla sibirica*. In: Pearson PL, Lewis KR (eds) Chromosomes today, Vol 5. Wiley, New York, pp 185–192

Weisblum B, de Haseth PL (1972) Quinacrine, a chromosome stain specific for deoxyadenylate-deoxythymidylate-rich regions in DNA. Proc Natl Acad Sci USA 69: 629–632

Yamasaki N (1956) Differentielle Färbung der somatischen Metaphasechromosomen von Cypripedium debile. Chromosoma 7: 620–626

Yunis JJ (1965) Interphase deoxyribonucleic acid condensation, late deoxyribonucleic acid replication, and gene inactivation. Nature 205: 311–312

VII
Chromosome Structural Aberrations

Origin of Structurally Abnormal Chromosomes

Chromosomes sometimes break spontaneously, or breakage may be caused by a mutagenic agent, such as ionizing radiation or a chemical compound. Unlike normal chromosome ends, broken ends tend to join each other. Usually the broken ends rejoin; in other words, the break heals. However, a break may lead to a deletion or, if more than one break has occurred in a cell, to structural rearrangements of chromosomes. At least three different DNA repair systems may be involved in the joining of broken chromosome ends.

Chromosomes may break at any stage of the cell cycle—G_1, S, G_2,—or mitosis, as well as during meiosis. Various cell types and stages show a very different response to chromosome breaking agents even in the same organism. Thus in *Vicia,* a dose of x-rays that causes approximately one aberration per cell in G_2 induces one aberration visible at metaphase per 10 cells when irradiation is given in G_1 (cf. Evans, 1974). Even greater differences are found when different plant and animal species are compared. When higher plants were tested for the dose of radiation needed to inhibit their growth or kill them, they showed more than 100-fold differences (Sparrow, 1965).

The study of chromosome breaks is intimately connected with research on gene mutations, because most mutagens induce both types of changes in the genetic material. Often the same agents are also carcinogenic. We know that the radiations and drugs used in cancer therapy can also cause malignant disease. Various aspects of chromosome breakage are reviewed in numerous books and articles, of which the following reflect a somewhat arbitrary sample: Kihlman, 1966; Rieger and Michaelis, 1967; Gebhart, 1970; Evans, 1962, 1974; Auerbach, 1976.

Chromosome Breaks and Rearrangements

If a chromosome breaks during the G_1 stage when it consists of only one chromatid, the break will be perpetuated in S and will affect *both* chromatids in the following metaphase. A single break may either rejoin or result in a deleted chromosome and an acentric fragment (Fig. VII.1b), which will be lost in a subsequent mitosis, or the acentric fragment is included in a daughter nucleus and replicates, which results in double fragments in the next metaphase.

Two breaks in the same chromosome may result in the formation of either a centric ring and an acentric fragment or an acentric ring and an

Fig. VII.1. Results of G_1 breaks in one chromosome (a), and in two chromosomes (f); (b) broken chromosome; (c) centric ring and acentric fragment; (d) acentric ring and centric fragment; (e) chromosome with pericentric inversion; (g) dicentric chromosome and acentric fragment; (h) balanced reciprocal translocation.

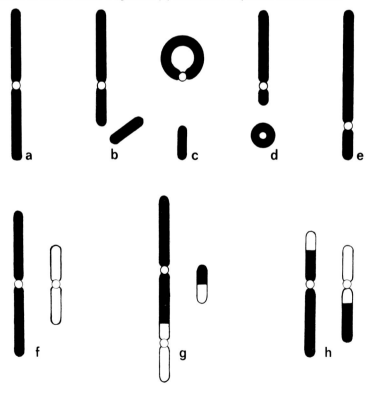

interstitial deletion (Figs. VII.1c and d and Fig. VII.2g). A segment that is deleted interstitially from a chromosome arm may remain as an acentric fragment if its ends fail to join. Very small fragments are called minutes.

If the breaks take place in the same arm, another possible result of a two-break intrachange is a paracentric inversion, in which the deleted segment rejoins the chromosome in an inverted position. If one break occurs in each arm, a pericentric inversion will be formed if the deleted

Fig. VII.2. Chromosome structural abnormalities. (a) Gap; (b) gap through centromere; (c) normal chromosome 1 and its homolog with a pericentric inversion; (d) dicentric chromosome and acentric fragment; (e) two dicentrics and two acentrics from the same cell (courtesy of EM Kuhn); (f) an interphasic D chromosome in satellite association with a normal D; (g) chromosome 9, ring(9) and double ring(9) (courtesy of ML Motl); (h) mitotic chiasma between heteromorphic homologs; (i) class II quadriradial between two D chromosomes in satellite association with a D and a G; (j–k) class IVb chromatid translocations; (l) hexaradial chromatid translocation or a satellite association between two D chromosomes and a G.

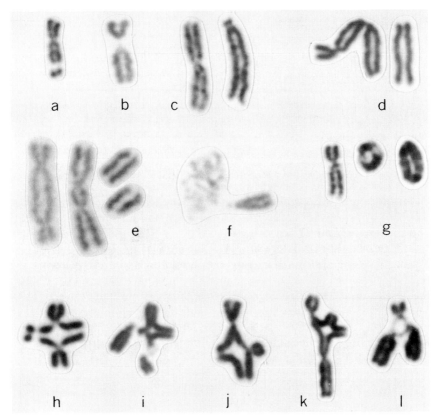

segment rejoins in an inverted position. (Figs. VII.1e and VII.2c). The latter rearrangement often results in the shifting of the position of the centromere. Many pericentric inversions and all paracentric inversions would be undetectable without banding techniques.

Naturally the first prerequisite for an interchange between two chromosomes is a break in each of them. For the broken ends to fuse they should not be too far from each other. The period during which broken ends are able to join also seems to be limited. An interchange may result in a reciprocal translocation (Fig. VII.1h) or a dicentric chromosome and an acentric fragment (Figs. VII.1g and VII.2d and e). The latter rearrangement is not stable because the acentric fragment will eventually be lost, and the dicentric will get into trouble sooner or later if the two centromeres are far enough apart for a twist to occur between them.

Multiple breaks in a cell may lead to several reciprocal translocations or other more complicated rearrangements, such as chromosomes with several centromeres. These chromosomes, with more than one centromere, usually do not survive more than one mitosis.

Chromatid Breaks and Rearrangements

When a break takes place during G_2, it involves as a rule only one of the two chromatids. A single break yields a deleted chromatid and an acentric fragment. Some of the possible consequences of chromatid breaks in two chromosomes are shown in Fig. VII.2i–k. Such configurations, which result from chromatid exchanges between two chromosomes, are called quadriradials. They come in two types, depending on whether the centromeres are on opposite sides (alternate) (I, IIIa, and IVa in Fig. VII.3), or next to each other (adjacent) (II, IIIb, and IVb in Fig. VII.3). The *alternate* type leads to the formation of two chromatids, each with a reciprocal translocation and two unchanged chromatids. The *adjacent* arrangement gives rise to a dicentric and an acentric chromatid and two unchanged chromatids.

Mitotic chiasmata (I in Fig. VII.3) form a special subgroup of the alternate type quadriradials. Here two homologous chromosomes have exchanged segments at corresponding points. The structure of a mitotic chiasma is similar to that of a meiotic bivalent with one chiasma (Figs. VII.2h and VII.5d). This implies that mitotic pairing and crossing-over have taken place in the segments involved. Thus a mitotic chiasma requires pairing of homologous segments and is different from random chromatid exchanges (Therman and Kuhn, 1976). For two alleles distal to the chiasma, mitotic crossing-over leads to somatic segregation (German, 1964). Various mosaicisms in rodents, including twin color spots in their fur, can best be explained as a result of mitotic crossing-over (cf. Grüneberg, 1966).

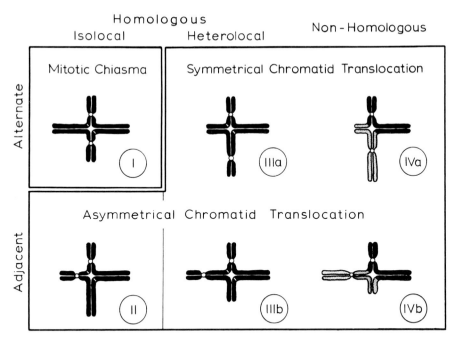

Fig. VII.3. The classification of quadriradial configurations (Therman and Kuhn, 1976).

Other alternate quadriradials also show mitotic segregation. The descendants of a cell in which a chromatid exchange has occurred are different from *each other,* whereas the descendants of a cell in which a chromosome aberration has taken place are identical. Many human mosaics display cell lines, with structurally different chromosomes, that owe their origin to segregation in a quadriradial (cf. Daly et al, 1977).

Telomeres

Unbroken chromosome ends do not as a rule show any tendency to join each other. They are assumed to be capped by so-called telomeres. A long-standing dispute is: Can a broken chromosome end heal and thereafter act as a telomere, or is a "real" telomere needed to make the end normal again?

There is evidence, at least in man, that broken chromosome ends are able to heal (Patau, 1965). (1) If a chromosome is broken with x-rays, the sister chromatids do not show a noticeable tendency to join, unlike the behavior of, say, plant chromosomes. (2) Cases of deleted chromosomes are too frequent and the breakpoints too consistent for them to be the result of two breaks. (3) In the cri du chat syndrome, which is caused by the deletion of about one-half of the short arm of chromosome 5, the broken ends may appear fuzzy, thus differing from normal chromosome ends (Patau, 1965). In addition, Niebuhr (1978) showed that in 35 cri du chat patients, the deletion appeared terminal in 27 of them, interstitial in four, and capped by a reciprocal translocation in four others. (4) Ring chromosomes may sometimes open up and act as normal two-armed chromosomes (for example, Cooke and Gordon, 1965).

However, it was recently shown with NOR-staining in two cases of translocation between the short arms of acrocentrics and other autosomes that the nucleolar organizing region of the short arm was attached to the broken end of the other translocation chromosome (Hansmann et al, 1977). Similarly, in two human meningiomas the broken ends of chromosomes 1 and 16 were capped by NOR-stained nucleolar organizer material (Zankl and Huwer, 1978). Such observations indicate that telomeres may be needed to cap the broken ends, at least for some breakpoints.

Telomere Association

Dutrillaux et al (1977) described a very interesting phenomenon in the Thiberge-Weissenbach syndrome. Unbroken chromosome ends tend to join each other, leading to the formation of chains and rings, sometimes involving all the chromosomes. No acentric fragments were present, indicating that this phenomenon cannot be the result of reciprocal translocations.

It is possible that in other diseases too, for instance in ataxia-telangiectasia, a similar although less extreme, end-to-end attachment of the chromosomes takes place (Hayashi and Schmid, 1975).

Triradial and Multiradial Chromosomes

Triradial and multiradial chromosome configurations deserve only brief mention because of their extreme rarity and are described here only as examples of the odd ways in which chromosomes may behave. In a triradial, part of the chromosome is duplicated; it has three branches in contrast to a quadriradial, which has four. Triradials are much rarer than quadriradials; for instance, Stahl-Maugé et al (1978) found two triradials in 53,000 cells of normal persons. However, they are more frequent in cells of patients with Fanconi's anemia and Bloom's syndrome and in cells treated with chromosome-breaking substances.

A triradial may—although rarely—come about through partial endoreduplication in that a part of the chromosome replicates twice, while the rest has replicated only once. Such a triradial of chromosome 1 is shown in Fig. VII.5e. Other types of triradials are relatively more common. They may be caused by a chromatid break in a previous mitosis and the adherence of the fragment to its sister chromatid. Another possible cause is a homologous triradial chromatid translocation (cf. Stahl-Maugé et al, 1978). The branching point in a triradial is often a fragile chromosome region as described in Chapter VIII.

Multibranched chromosomes have so far been found only in the lymphocytes of one patient who was deficient for IgA and IgE (cf. Tiepolo et al, 1979). These peculiar configurations consist of variable numbers and combinations of short and long arms of chromosomes 1, 9, and 16 in which the exchanges occur at the centric regions.

Centric Fusion

Whole-arm transfers, or Robertsonian translocations as they also are called, constitute a special class of reciprocal translocations. In man they almost always occur between two acrocentric chromosomes and are the most commonly observed chromosome aberrations.

The origin of Robertsonian translocations has often been assumed to be the fusion of two centromeres that have each broken in the middle. Recent banding studies show that this is not the only, and possibly not even the most common, mechanism. As illustrated in Fig. VII.4, depending on whether the break in the two acrocentrics occurs through the centromere or on the short or the long arm side of it, the result is a monocentric or dicentric (or acentric) translocation chromosome, consisting of two long arms. The reciprocal product, made up of two short

Fig. VII.4. Origin of Robertsonian translocations between a G and a D chromosome. Left: Breaks on the short arms resulting in a dicentric and acentric chromosome; Right: break either through the centromeres or one break on the short arm, the other on the long arm resulting in two monocentric chromosomes.

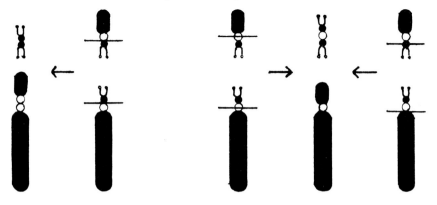

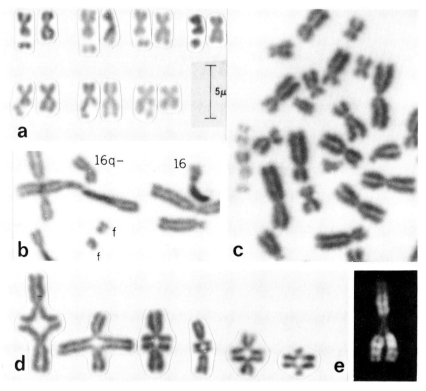

Fig. VII.5. (a) Chromosome 16 with a fragile region and its normal homolog from 7 cells; (b) break at the fragile region and the replicated fragment (f); (c) banded C chromosome left behind in its cycle; (d) mitotic chiasmata involving different human chromosome pairs; (e) triradial chromosome 1 resulting from partial endoreduplication.

arms, is inevitably lost if it does not have a centromere, In fact, even when such a chromosome possesses a centromere, it is usually lost because it consists entirely of heterochromatin. Neither the absence of these small chromosomes nor their presence as extra units seems to affect the phenotype.

Inactivation of the Centromere

In dicentric human chromosomes the second centromere is turned off in many cases. The chromosome functions as a normal monocentric chromosome would (cf. Therman et al, 1974). No second primary constriction is visible, but a C-band marks its position (Fig. VI.1g). Whether all human chromosomes have this ability and how the inactivation process works, are still unanswered questions. Inactivated centromeres have been found in dicentric Robertsonian translocation chromosomes between two acrocentrics but also in a number of other chromosomes (cf. Pallister et al, 1974). This phenomenon has been studied especially in various X-X translocations (cf. Maraschio et al, 1977).

A somewhat similar phenomenon is found in monocentric human X chromosomes. They have the appearance of acentric fragments, but a C-band is present and they have the ability to divide, although less effectively than normal chromosomes. As a result, some cells lack an X chromosome, whereas others contain several of these abnormal chromosomes. This may be the mechanism that accounts for the increased frequency of cells with a 45,X chromosome constitution in older women (cf. Fitzgerald and McEwan, 1977).

Misdivision

Misdivision, which will be discussed in connection with meiosis but which may also occur in mitotic anaphases, implies a transverse rather than a longitudinal division of the centromere (Fig. XI.2). This leads to the formation of *isochromosomes* and *telocentrics*.

Interphase-like Chromosomes in Metaphase

When two cells, each in a different phase of its cycle, fuse, the nucleus that is in mitosis compels the chromosomes of the interphase nucleus to condense. This may also happen to a micronucleus in a mitotic cell. Such chromosomes have been called prematurely condensed (PCC) (cf. Rao, 1977). Depending on which phase the interphase nucleus is in, such chromosomes will appear as long single chromatids, interphase-like (these have also been called pulverized), or banded and stretched. This phenomenon allows us to study interphase chromosomes, which otherwise would be impossible.

One or a few chromosomes that are still in interphase or stretched out when the rest of the chromosome complement is in metaphase are often found after a cell is treated with a mutagen. They are observed only rarely in untreated cells. Sometimes only one arm of the chromosome is affected. The interphase-like chromosomes may be in the middle of DNA synthesis. Although some such chromosomes may originate from micronuclei, most of them have probably been left behind in their cycle. For instance, one-half of an interphase-like chromosome or an affected acrocentric that is nose-to-nose with other G and D chromosomes (Fig.

VII.2f) can hardly have come from a micronucleus. It seems therefore singularly inappropriate to call such chromosomes prematurely condensed, when they are actually lagging in their cycle. Such a chromosome (possibly 11) is shown in Fig. VII.5c.

Because these out-of-phase chromosomes, whatever their origin, are left behind when the normal chromosomes divide, aneuploidy is the natural result. Whether they ever give rise to structurally aberrant chromosomes, for instance when only part of the chromosome is affected, is not known.

Sister Chromatid Exchanges

It has been known for 20 years that sister chromatids sometimes exchange homologous segments. However, since the methods, such as autoradiography or the more recent BrdU technique (Latt, 1973), that are used to reveal this phenomenon also induce it, its frequency in untreated cells remains a matter of extrapolation (cf. Wolff, 1977). In Chinese hamster cells the estimate varies from 2.3 to 5 breaks per cell per two cell generations (cf. Wolff, 1977). Figure V.2 illustrates a human lymphocyte in which the sister chromatid exchanges have been made visible by coriphosphine O and a Chinese hamster ovary cell stained using the Giemsa technique of Korenberg and Freedlender (1974).

Sister chromatid exchanges provide a much more sensitive system for analyzing mutagen action than do chromosome or chromatid breaks (cf. Wolff, 1977). Considerable increases in sister chromatid exchanges are induced by mutagens, such as alkylating agents. However, increases in chromosome breakage and in sister chromatid exchanges do not necessarily go hand in hand. For example, in Bloom's syndrome both types of events are greatly increased, whereas in Fanconi's anemia, in which up to 20 percent of lymphocytes may contain chromosome breaks, no increase in sister chromatid exchanges is apparent (cf. Wolff, 1977).

References

Auerbach C (1976) Mutation research. Problems, results and perspectives. Chapman and Hall, London

Auerbach C (1978) Forty years of mutation research: a pilgrim's progress. Heredity 40: 177–187

Cooke P, Gordon RR (1965) Cytological studies on a human ring chromosome. Ann Hum Genet 29: 147–150

Daly RF, Patau K, Therman E, et al (1977) Structure and Barr body formation of an Xp+ chromosome with two inactivation centers. Am J Hum Genet 29: 83–93

Dutrillaux B, Aurias A, Couturier J et al (1977) Multiple telomeric fusions and chain configurations in human somatic chromosomes. In: de la Chapelle A, Sorsa M (eds) Chromosomes today, Vol 6. Elsevier/North-Holland, Amsterdam, pp 37–44

Evans HJ (1962) Chromosome aberrations induced by ionizing radiations. Int Rev Cytol 13: 221–321

Evans HJ (1974) Effects of ionizing radiation on mammalian chromosomes. In: German J (ed) Chromosomes and cancer. Wiley, New York pp 191–237

Fitzgerald PH, McEwan CM (1977) Total aneuploidy and age-related sex chromosome aneuploidy in cultured lymphocytes of normal men and women. Hum Genet 39: 329–337

Gebhart E (1970) The treatment of human chromosomes in vitro: results. In: Vogel F, Röhrborn G (eds) Chemical mutagenesis in mammals and man. Springer, New York pp 367–382

German J (1964) Cytological evidence for crossing-over in vitro in human lymphoid cells. Science 144: 298–301

Grüneberg H (1966) The case for somatic crossing over in the mouse. Genet Res 7: 58–75

Hansmann I, Wiedeking C, Grimm T, et al (1977) Reciprocal or nonreciprocal human chromosome translocations? The identification of reciprocal translocations by silver staining. Hum Genet 38: 1–5

Hayashi K, Schmid W (1975) Tandem duplication q14 and dicentric formation by end-to-end chromosome fusions in Ataxia telangiectasia (AT). Clinical and cytogenetic findings in 5 patients. Humangenetik 30: 135–141

Korenberg JR, Freedlender EF (1974) Giemsa technique for the detection of sister chromatid exchanges. Chromosoma 48: 355–360

Kihlman BA (1966) Actions of chemicals on dividing cells. Prentice-Hall, Englewood Cliffs, New Jersey

Latt SA (1973) Microfluorometric detection of deoxyribonucleic acid replication in human metaphase chromosomes. Proc Natl Acad Sci USA 70: 3395–3399

Maraschio P, Scappaticci S, Ferrari E, et al (1977) X chromosomes attached by their long arm: replication autonomy of the short arm adjacent to the inactive centromere. Ann Genet 20: 179–183

Niebuhr E (1978) Cytologic observations in 35 individuals with a 5p-karyotype. Hum Genet 42: 143–156

Pallister PD, Patau K, Inhorn SL, et al (1974) A woman with multiple congenital anomalies, mental retardation and mosaicism for an unusual translocation chromosome t(6;19). Clin Genet 5: 188–195

Patau K (1965) The chromosomes. In: Birth defects original article series, Vol 1, No 2. The National Foundation - March of Dimes, New York, pp 71–74

Rao PN (1977) Premature chromosome condensation and the fine structure of chromosomes. In: Yunis JJ (ed) Molecular structure of human chromosomes. Academic, New York, pp 205–231

Rieger R, Michaelis A (1967) Die Chromosomenmutationen. Gustav Fischer, Jena

Sparrow AH (1965) Comparisons of the tolerances of higher plant species to acute and chronic exposures of ionizing radiation. In: Mechanisms of the dose rate effect of radiation at the genetic and cellular levels. Special suppl. Jpn J Genet 40: 12–37

Stahl-Maugé C, Hager HD, Schroeder TM (1978) The problem of partial endo-reduplication. Hum Genet 45: 51–62

Therman E, Kuhn EM (1976) Cytological demonstration of mitotic crossing-over in man. Cytogenet Cell Genet 17:254–267

Therman E, Sarto GE, Patau K (1974) Apparently isodicentric but functionally monocentric X chromosome in man. Am J Hum Genet 26: 83–92

Tiepolo L, Maraschio P, Gimelli G, et al (1979) Multibranched chromosomes 1, 9, and 16 in a patient with combined IgA and IgE deficiency. Hum Genet 51: 127–137

Wolff S (1977) Sister chromatid exchange. Ann Rev Genet 11: 183–201

Zankl H, Huwer H (1978) Are NORs easily translocated to deleted chromosomes? Hum Genet 42: 137–142

VIII
Causes of Chromosome Breaks

Spontaneous Chromosome Breaks

Spontaneously broken and rearranged chromosomes are occasionally found in every human being and every cell culture. Their frequency varies from person to person and from culture to culture. One rule, however, appears to be well established. The incidence of aneuploid cells is known to increase with age, and similarly chromosome structural changes become more frequent in older persons. In our laboratory an analysis of 2324 cells from persons under 40 years of age gave an average of 0.8 percent of cells with chromosome structural aberrations (excluding gaps), whereas in another sample of about the same size from individuals whose average age was 55.8 years, abnormalities were found in 2.4 percent of cells (Kuhn and Therman, 1979b). Interestingly, the sensitivity of lymphocytes to alkylating agents increases significantly from newborns to young adults (mean age 23) to old people (mean age 70) (Bochkov and Kuleshov, 1972).

Although the trends are the same, the actual frequencies of chromosome aberrations found in different laboratories show a wide range of variation. This probably depends mainly on different criteria being used for scoring aberrations and also on culture conditions.

Another indication of an age effect is the fact that tissue cultures of normal human fibroblasts can be grown unchanged through only about 50 passages (approximately one year) (Hayflick and Moorhead, 1961). Thereafter cell growth suffers and more and more aberrations appear. Finally the normal cell strains are transformed into permanent cell lines with numerically and structurally aberrant chromosome constitutions (cf. Nichols, 1975).

When we talk about ''spontaneous'' chromosome breaks, it means that we can only guess at the actual causes. The genotype of the individual undoubtedly determines the basic level of aberrations. Physiological degeneration in old age probably accounts for the increase in both nondisjunction and breakage of chromosomes. Exposure to cosmic rays and medical or occupational radiation constitute other sources of chromosome damage. Various drugs, viral infections, and even high fever may be yet other sources of damage.

It is well known that chemically induced chromosome breaks do not occur at random among cells; in other words they do not follow a Poisson distribution. The same seems to be true of spontaneous breaks (cf. Schroeder and German, 1974; Therman and Kuhn, 1976). Part of the evidence comes from individuals with complicated chromosome rearrangements and is thus incidental. In a population of chromosomally normal grasshoppers, Coleman (1947) found an individual who was heterozygous for three reciprocal translocations and one inversion. Similarly White (1963) described a grasshopper with a complex translocation involving breaks in four nonhomologous chromosomes. In man, too, complicated rearrangements are occasionally encountered. For example, in a child with congenital anomalies, four chromosomes showed altogether six breaks (Seabright et al, 1978), and Bijlsma et al (1978), who have reviewed the literature on multiple break cases, report a family in which two reciprocal translocations were segregating.

Points of Weakness in Human Chromosomes

A further cause for spontaneous chromosome breakage is found in certain persons. In such cases a specific chromosome region, which often appears as a nonstained and stretched-out gap, functions as a point of weakness.

The chromosome is likely to break at this region, although the incidence of breakage varies greatly from person to person (cf. Giraud et al, 1976). Such segments have been described in a number of chromosomes, in X, 2, 10, 11, 12, 17, 20, and especially in 16q (Fig. VII.5a and b) (cf. Hecht and Kaiser-McKaw, 1979). The differential "weak" segment on a specific chromosome always appears to be the same. It is interesting that the weak point on 2q seems to be the place where two chimpanzee chromosomes—or more accurately the chromosomes of a common primate ancestor—have joined to form one human chromosome.

Radiation-Induced Breaks

Ultraviolet light, and especially various types of ionizing radiation, are powerful chromosome-breaking agents. The effects of a wide range of radiations, such as x-rays, γ-radiation, α and β particles, and neutrons, have been studied in this respect. Ionizing radiation causes chromosome breaks at any stage of the mitotic cycle or during meiosis, although the vulnerability of different phases and different organisms varies greatly. The results also vary according to whether a given dose of radiation is applied within a short time span or over a longer period or is fractionated. The tritium in ^{3}H-thymidine, so widely used for autoradiography, emits β particles that may cause considerable damage when incorporated into the chromosomes.

An enormous amount of research has been undertaken to elucidate the actual mechanism of radiation-induced chromosome breakage. In the late 1930s the so-called target theory was born. It states that a chromosome is broken when it is hit directly by an ion or an ion cluster. However, because a number of observations cannot be explained on the basis of the target theory, the chemical theory of chromosome breakage was proposed in the 1940s. This theory says that most of the chromosome breakage is done by substances that result from the chemical reactions induced by radiation. The various radicals that arise from irradiated water molecules seem to be especially active in this respect.

The following are some of the observations that do not agree with the target theory but which can be readily explained by the chemical theory. Chromosomes of many plants are broken with much lower doses of radiation than, for instance, *Drosophila* chromosomes. Chromosome breaks can be caused by a preirradiated medium. One of the most important observations is that oxygen greatly enhances the effects of radiation, and that inversely a number of reducing substances— especially those containing sulfhydryl groups—counteracts radiation damage (cf. Evans, 1974). As has been stressed repeatedly, especially by Auerbach (1976, 1978), reactions caused by radiations—as well as those induced by chemical mutagens—take place in a *living cell,* a fact that in many ways influences the final outcome of any treatment.

Chemically Induced Breaks

The list of substances found to break chromosomes is already very long and is still growing steadily. It includes alkylating agents, nucleic acid analogues, purines, antibiotics, nitroso-compounds, and a large number of miscellaneous substances (cf. Kihlman, 1966; Rieger and Michaelis, 1967; Gebhart, 1970; Auerbach, 1976). Such substances have often been called radiomimetic or clastogenic.

Most chemicals induce breaks of the G_2 type, but they have to be present during the preceding S period. As with radiation treatment, different organisms and tissues show a wide range of responses to the same chemical. An example is illustrated in Fig. VIII.1. Two anticancer drugs of the methyl-benzyl hydrazine group broke no chromosomes

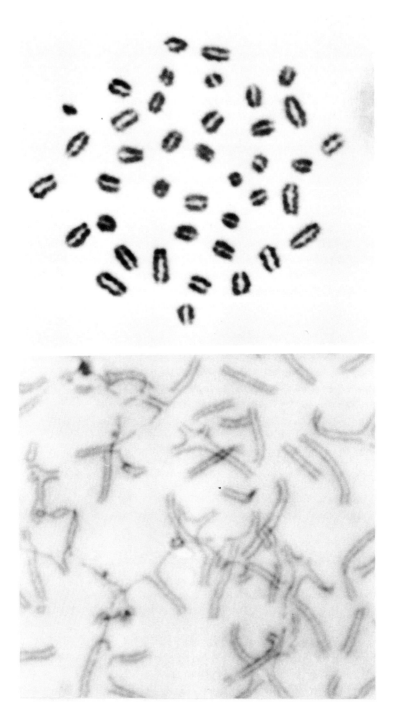

Fig. VIII.1. Top: Normal spermatogonial metaphase of the mouse ($2n = 40$). Bottom: Chromosome breakage in a mouse cancer cell caused with 1-methyl–2–benzyl hydrazine (Therman, 1972).

either in malignant or normal cells in vitro. In vivo they showed no effect on the chromosomes in normal mouse spleen, bone marrow, or spermatocytes. However, they caused extensive damage in transplantable ascites tumors in the same animal (Therman, 1972; Rapp and Therman, 1977). It seems obvious that the mouse is needed to metabolize and change the drug and that it undergoes a second change in the cancer cells. Apparently at least two steps are needed to transform the methyl-benzyl hydrazines into chromosome-breaking substances. All the breaks are of the G_2 type and take place in the Q-dark or G-light regions (Rapp and Therman, 1977).

Virus-Induced Breaks

The discovery that many viruses induce chromosome breaks is relatively recent. This aspect of chromosome breakage has therefore been studied much less than has radiation or chemical mutagenesis, and very little is known about the interaction between the virus and the chromosomes that is required to bring about chromosome breaks. Viral chromosome breaks may or may not undergo repair. Harnden (1974b; Nichols, 1975) has reviewed the information on chromosome-breaking viruses.

Chromosome damage caused by viral infection varies from single chromosome and chromatid breaks to rearrangements and total pulverization of the chromosome complement. The latter is found especially in lymphocytes of persons with an acute viral infection, such as measles. What proportion, if any, of the spontaneous chromosome aberrations in man is attributable to viral action is still unknown.

Genetic Causes of Chromosome Breakage

As already pointed out, different persons may show consistently different rates of spontaneous chromosome breakage, probably depending on their genetic makeup. In a number of syndromes, most of which are caused by single recessive autosomal genes, the incidence of chromosome breaks and rearrangements is greatly increased. This phenomenon has been analyzed mainly in cultured lymphocytes, but fibroblasts and bone marrow cells are also affected, although to a lesser degree. The homozygous individuals also have a greatly increased probability of developing malignant disease (cf. Schroeder and Kurth, 1971; German, 1974).

The first syndrome in which this type of chromosome instability was discovered was Fanconi's anemia (Schroeder et al, 1964). The patients usually die as children due to anemia or leukemia. Chromosome breakage takes place almost exclusively in G_2 and leads mainly to the formation of quadriradials between nonhomologous chromosomes (Figs. VII.2i–k and

VII.3). This condition is characterized by stunted growth and other congenital anomalies, but especially the development of aplastic anemia (deficiency of all elements in the blood).

Another condition that is fascinating from a cytogeneticist's point of view is the rare Bloom's syndrome. This disease is characterized by pronounced stunted growth. Other symptoms are: a characteristic face, sensitivity to sunlight, and telangiectactic (dilation of small vessels and arteries) erythema (cf. German, 1974). Chromosome breaks seem to take place mainly in S-G_2, leading to a wide variety of aberrations (Kuhn and Therman, 1979a).

The most interesting feature of chromosome aberrations in Bloom's syndrome is the high frequency of mitotic chiasmata, which make up 90 percent of the quadriradials (Figs. VII.2h, VII.3, and VII.5d). Up to 15 percent of the lymphocytes may contain a chiasma as compared to 1/1000 cells in normal persons (Therman and Kuhn, 1976). The tendency to malignant disease is even greater than in Fanconi's anemia (cf. Schroeder and Kurth, 1971; German, 1974). A further although less studied disease showing spontaneous chromosome breakage is ataxia-telangiectasia (cf. Harnden, 1974a).

Nonrandomness of Chromosome Breaks

The nonrandomness of chromosome breaks may be observed in different cells or various chromosome segments within a cell. As previously mentioned, a nonrandom distribution between cells applies to spontaneous breaks as well as to breaks induced by Fanconi's anemia, Bloom's syndrome, and chemical compounds.

Banding techniques reveal that breaks in human chromosomes, whatever their cause, practically always take place in the Q-dark (G-light) chromosome regions. However, it is not clear whether breaks occur preferentially in these segments or whether the nonrandomness is the result of differential repair. Breaks are very unevenly distributed even among the Q-dark segments, with certain of them emerging as clear hot spots. But the hot spots are not the same when chromosome breakage caused by different agents is compared (cf. von Koskull and Aula, 1977).

An extensive study on the localization of exchange points in mitotic chiasmata, characteristic of Bloom's syndrome, was made by Kuhn (1976). She found that the overwhelming majority of cross-over points was localized in the Q-dark regions, with a few of them standing out as definite hot spots.

Studies on the effects of chemical mutagens show that the breaks are often concentrated in heterochromatic regions of chromosomes. To take one much-studied example, mitomycin C induces breaks in human chromosomes largely in the constitutive heterochromatin of chromosomes 1, 9, 16, and the acrocentrics (Shaw and Cohen, 1965).

Methods in Chromosome-Breakage Studies

In plants and animals, chromosome breakage has been analyzed both in vivo and in vitro. In-vivo studies are naturally preferable for many investigations, since they permit us to observe the effects of a mutagen in an intact organism. Effects in vivo often differ drastically from the mutagen's effects on cells in culture.

In man, in-vivo experiments are usually impractical or unethical, or both. Nevertheless it is possible to analyze chromosome damage in persons who have been exposed to ionizing radiation for medical reasons, because of their occupation, or as a result of an accident. The most extensively studied group is the atomic bomb survivors of Hiroshima and Nagasaki. The results of chromosome studies done on them are reviewed by Awa (1974). The victims show a significantly higher incidence of chromosome abnormalities in their lymphocytes than do nonexposed controls. One of the most interesting phenomena is that those who received heavy doses of radiation still exhibit, more than 20 years later, abnormal chromosomes, such as dicentrics, acentrics, reciprocal translocations, inversions, rings, and deleted chromosomes. Often clones of cells with the same abnormal chromosome constitution have been found. The same is true of persons who have been exposed to great amounts of radiation for medical reasons.

Gebhart (1970) has compared the effects of various drugs, most of them used in cancer chemotherapy, on cells of treated persons and on lymphocytes exposed in vitro.

However, most of the information we have about the induction of chromosome aberrations has been obtained from studies on cultured cells, both lymphocytes and fibroblasts. Cultured human cancer cells, especially various strains of HeLa cells (long-term human cell lines originating in a cervical carcinoma), have also been used in mutagenesis experiments. Numerous studies on the effects of chromosome-breaking agents have been performed on transplantable ascites tumors of the mouse, which are grown in a liquid form in the abdominal cavity of the animal (cf. Adler, 1970). This method makes it easy to obtain beautiful chromosome preparations.

Different methods have been employed to determine chromosome damage (cf. Schoeller and Wolf, 1970; Nichols, 1973). One of the earlier techniques consists of scoring dicentric bridges and acentric fragments in anaphase. Another method involves counting micronuclei formed by damaged or lagging chromosomes in mitoses following the treatment (cf. Nichols, 1973).

The most accurate results are naturally achieved by analyzing whole metaphase plates, especially by the various banding techniques. Even when this is done, however, the results of comparable studies conducted by different groups often yield discordant results. Apart from variation in the biological material and in culture techniques, the discrepancies depend mainly on different criteria used in scoring chromosome and chromatid aberrations. For example, one frequent source of confusion is the scoring of so-called gaps. These are mostly despiralized chromosome regions that appear as thinner and less stained regions in the chromatids. Since these gaps are difficult to delimit in either direction and as a rule do not lead to permanent chromosome damage, it would be best not to score them at all. Furthermore, since even a broken chromatid fragment tends to stick to its sister chromatid, it is often impossible to decide whether or not a true break has taken place. The dislocation of a fragment has commonly been used as a criterion to distinguish breaks from gaps (Schoeller and Wolf, 1970). Nevertheless many scored breaks may, in reality, represent gaps. In addition, in most studies no definition or illustration is furnished to show according to what criteria the various aberrations are classified.

In view of these difficulties, the most accurate procedure would be to score only dicentrics and quadriradials, since there can hardly be any disagreement about them. However, since these configurations require two breaks, they are much rarer than one-break aberrations. Consequently, to collect an adequate number of them often requires that a prohibitive number of cells be checked, or that the concentration of a drug or the amount of radiation used be so high as to interfere with cell division or cause cell death. Consequently many recent studies on chromosome-breaking mutagens take advantage of the more sensitive system of sister chromatid exchanges, which is free from many of the difficulties just discussed. However, it should be kept in mind that the correlation of chromosome breaks and sister chromatid exchanges is far from perfect.

Whatever criteria are used for the analysis of chromosome aberrations, it is important that control cells from the same persons be cultured simultaneously. The analysis of the treated and control cells should be done on randomized coded slides to avoid systematic errors and the bias of the investigator (cf. Schoeller and Wolf, 1970; Cohen and Hirschhorn, 1971).

References

Adler I-D (1970) Cytogenetic analysis of ascites tumour cells of mice in mutation research. In: Vogel F, Röhrborn G (eds) Chemical mutagenesis in mammals and man. Springer, New York, pp 251–259

Auerbach C (1976) Mutation research. Problems, results and perspectives. Chapman and Hall, London

Auerbach C (1978) Forty years of mutation research: a pilgrim's progress. Heredity 40: 177–187

Awa AA (1974) Cytogenetic and oncogenic effects of the ionizing radiations of the atomic bombs. In: German J (ed) Chromosomes and cancer. Wiley, New York, pp 637–674

Bijlsma JB, deFrance HF, Bleeker-Wagenmakers LM, et al (1978) Double translocation t(7;12),t(2;6) heterozygosity in one family. A contribution to the trisomy 12p syndrome. Hum Genet 40: 135–137

Bochkov NP, Kuleshov NP (1972) Age sensitivity of human chromosomes to alkylating agents. Mutation Res 14: 345–353

Cohen MM, Hirschhorn K (1971) Cytogenetic studies in animals. In: Hollaender A (ed) Chemical mutagens. Principles and methods for their detection, Vol 2. Plenum, New York, pp 515–534

Coleman LC (1947) Chromosome abnormalities in an individual of *Chorthippus longicornis* (Acrididae). Genetics 32: 435–447

Evans HJ (1974) Effects of ionizing radiation on mammalian chromosomes. In: German J (ed) Chromosomes and cancer. Wiley, New York, pp 191–237

Gebhart E (1970) The treatment of human chromosomes in vitro: results. In: Vogel F, Röhrborn G (eds) Chemical mutagenesis in mammals and man. Springer, New York, pp 367–382

German J (1974) Bloom's syndrome. II. The prototype of human genetic disorders predisposing to chromosome instability and cancer. In: German J (ed) Chromosomes and cancer. Wiley, New York, pp 601–617

Giraud F, Ayme S, Mattei JF, et al (1976) Constitutional chromosomal breakage. Hum Genet 34: 125–136

Harnden DG (1974a) Ataxia telangiectasia syndrome: cytogenetic and cancer aspects. In: German J (ed) Chromosomes and cancer. Wiley, New York, pp 619–636

Harnden DG (1974b) Viruses, chromosomes, and tumors: the interaction between viruses and chromosomes. In: German J (ed) Chromosomes and cancer. Wiley, New York, pp 151-190

Hayflick L, Moorhead PS (1961) The serial cultivation of human diploid cell strains. Exp Cell Res 25: 585–621

Hecht F, Kaiser-McKaw B (1979) The importance of being a fragile site. Am J Hum Genet 31: 223–225

Kihlman BA (1966) Actions of chemicals on dividing cells. Prentice-Hall, Englewood Cliffs, New Jersey

Koskull H von, Aula P (1977) Distribution of chromosome breaks in measles, Fanconi's anemia and controls. Hereditas 87: 1–10

Kuhn EM (1976) Localization by Q-banding of mitotic chiasmata in cases of Blooms's syndrome. Chromosoma 57: 1–11

Kuhn EM, Therman E (1979a) Chromosome breakage and rejoining of sister chromatids in Bloom's syndrome. Chromosoma 73: 275–286

Kuhn EM, Therman E (1979b) No increased chromosome breakage in three Bloom's syndrome heterozygotes. J Med Genet 16: 219–222

Nichols WW (1973) Cytogenetic techniques in mutagenicity testing. Agents and Actions 3: 86–92

Nichols WW (1975) Somatic mutation in biologic research. Hereditas 81: 225–236

Rapp M, Therman E (1977) The effect of procarbazine on the chromosomes of normal and malignant mouse cells. Ann Genet 20: 249–254

Rieger R, Michaelis A (1967) Die Chromosomenmutationen. Gustav Fischer, Jena

Schoeller L, Wolf U (1970) Possibilities and limitations of chromosome treatment in vitro for the problem of chemical mutagenesis. In: Vogel F, Röhrborn G (eds) Chemical mutagenesis in mammals and man. Springer, New York, pp 232–240

Schroeder TM, Anschütz F, Knopp A (1964) Spontane Chromosomenaberrationen bei familiärer Panmyelopathie. Humangenetik 1: 194–196

Schroeder TM, German J (1974) Bloom's syndrome and Fanconi's anemia: demonstration of two distinctive patterns of chromosome disruption and rearrangement. Humangenetik 25: 299–306

Schroeder TM, Kurth R (1971) Spontaneous chromosomal breakage and high incidence of leukemia in inherited disease. Blood 37: 96–112

Seabright M, Gregson N, Pacifico E, et al (1978) Rearrangements involving four chromosomes in a child with congenital abnormalities. Cytogenet Cell Genet 20: 150–154

Shaw MW, Cohen MM (1965) Chromosome exchanges in human leukocytes induced by mitomycin C. Genetics 51: 181–190

Therman E (1972) Chromosome breakage by 1-methyl–2–benzylhydrazine in mouse cancer cells. Cancer Res 32: 1133–1136

Therman E, Kuhn EM (1976) Cytological demonstration of mitotic crossing-over in man. Cytogenet Cell Genet 17: 254–267

White MJD (1963) Cytogenetics of the grasshopper *Moraba scurra* VIII. A complex spontaneous translocation. Chromosoma 14: 140–145

IX
The Main Features of Meiosis

Significance of Meiosis

The most important modification of mitosis is meiosis, which is the reduction division that gives rise to the haploid generation in the life cycle. In mammals, the haploid generation is restricted to one cell, the gamete, whereas the other cells in the animal itself are diploid. However, in many other organisms, especially lower plants, the haploid generation is more important than the diploid; or the two generations may be more or less equal, as in the mosses. In those organisms in which the haploid phase dominates, the diploid generation may be represented by only one cell, the zygote. Under those circumstances, meiosis takes place immediately after fertilization.

In the alternation of the haploid and diploid generations, the main events are fertilization, which doubles the chromosome number, and meiosis, which halves it. The reduction of the chromosome number during meiosis occurs because the nucleus divides twice, while the chromosomes replicate only once. Another essential characteristic of meiosis is the pairing of homologous chromosomes, which makes their orderly segregation possible.

A further phenomenon, typical of meiosis in most organisms, is the crossing-over or exchange of homologous segments between two of the four chromatids of the paired chromosomes. Crossing-over as such is not a prerequisite for orderly meiotic segregation. For instance, in the *Drosophila* male normal meiosis occurs, but crossing-over is absent. Another indication of the mutual independence of crossing-over and segregation is provided by *mitotic* crossing-over, which does not result in segregation of homologous chromosomes.

If crossing-over is absent in one sex, it is usually present in the other. Thus in the *Drosophila* female, crossing-over takes place regularly; its failure leads to the dissociation of paired chromosomes and resultant abnormalities in segregation. As a rule, in organisms in which crossing-over is established, it is a prerequisite for orderly chromosome segregation.

The almost ubiquitous presence of crossing-over, in at least one sex per species, carries obvious evolutionary benefits. In the offspring, it greatly increases genetic recombination beyond the variability already derived from the independent segregation of maternal and paternal chromosomes in the first meiotic anaphase.

The main features of meiosis, namely: one DNA synthesis, two cell divisions, chromosome pairing, crossing-over, and segregation are strikingly similar throughout the plant and animal kingdoms. As Darlington puts it: "The lily can tell us what happens in the mouse. The fly can tell us what happens in man." (Darlington and La Cour, 1976, p. 20).

Meiotic Stages

We will follow the premeiotic and meiotic stages in the pollen mother cells of a liliaceous plant, *Eremurus*, which has 14 relatively large chromosomes (Oksala and Therman, 1958; Therman and Sarto, 1977). Before meiosis, the pollen mother cells, which are located in the anther lobes, undergo a number of mitoses. The last of these is shown in Fig. IX.1. Despite a number of claims that the homologous chromosomes show a tendency toward mitotic pairing during the premeiotic stages, no sign of the homologous chromosomes lying side by side is observed. It seems that little, if any, solid evidence exists for chromosome pairing before meiosis (cf. Walters, 1970; John, 1976).

Premeiotic Interphase

Premeiotic contraction has been described in a number of organisms, especially plants (cf. Walters, 1970). Toward the end of the last premeiotic interphase, the chromosomes contract and come to resemble mitotic prophase chromosomes. Later in interphase these contracted chromosomes unravel to form the leptotene threads. No sign of such premeiotic contraction is seen in *Eremurus* (Fig. IX.1). Indeed this phenomenon varies greatly between closely related species and even between cells in the same plant. Possibly these contracted chromosomes correspond to the so-called prochromosomes observed in the meiocytes of many animals, including human oocytes.

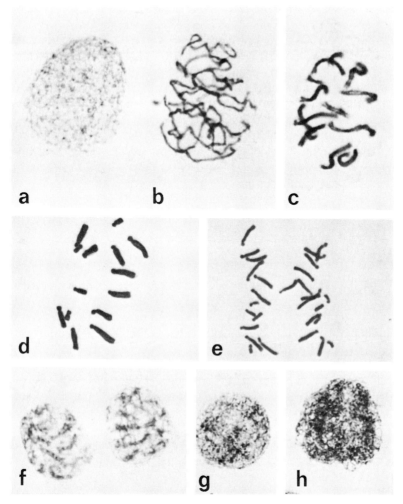

Fig. IX.1. Premeiotic stages and leptotene in pollen mother cells of *Eremurus*. (a) Interphase; (b–c) prophase; (d) metaphase; (e) anaphase; (f) telophase; (g) interphase; (h) leptotene (Feulgen squash).

In some organisms, premeiotic DNA synthesis takes place in the interphase preceding leptotene (cf. Stern and Hotta, 1974). In the liliaceous plant *Trillium,* the period between DNA synthesis and the beginning of meiosis is a couple of weeks, in *Lilium* a few days. However, in *Eremurus* and in human oocytes, this synthesis occurs in early leptotene (Therman and Sarto, 1977).

Leptotene (or Leptonema)

The beginning of meiotic prophase is characterized by the chromosomes becoming visible as thin threads (Fig. IX.1h). Thicker points, or chromomeres, which are coiled segments in the stretched-out chromosomal strand, appear and become clearer as the chromosomes gradually contract. At this stage in *Eremurus,* DNA synthesis can be demonstrated by autoradiography. An ephemeral stage that has been called the "distance pairing" stage follows after typical leptotene (Therman and Sarto, 1977). During this phase the homologues lie side by side for long stretches, apparently without touching. It is possible that the homologous chromosomes sort themselves out in some way during this stage (Fig. IX.2a); or it may represent early zygotene; synapsis may have started at some points, although it is not visible.

Zygotene (or Zygonema)

In zygotene, the chromosome ends collect at a spot inside the nuclear membrane close to the centrioles, which are outside the nucleus. This movement causes the so-called bouquet formation (Fig. IX.3). Although the bouquet configuration varies, it seems to be ubiquitous in both plants and animals (Oksala and Therman, 1958). It may be one of the mechanisms that allow the homologous chromosomes to find each other.

The homologous chromosomes pair during zygotene. Parts of the chromosomes are thin leptotene threads, whereas the rest have synapsed to form thicker pachytene-like chromosomes. The points where the pairing has started can be clearly seen (Fig. IX.4). Now the chromomeres are visible as different-sized "beads" in the chromosome thread.

That paired and unpaired regions are visible in the same chromosome pair indicates that the pairing of homologues starts at several points. Synapsis takes place not only between two homologous chromosomes but between strictly homologous segments. This is convincingly demonstrated by the pairing configuration of two homologues, if one of them has an inversion. To achieve point-by-point pairing in the inverted segment, one of the chromosomes has to form a loop, and such loops are regularly seen. Similarly, if a translocation has taken place between two chromosomes, the corresponding segments of translocated and nontranslocated chromosomes synapse, leading to a characteristic X-shaped figure.

Pachytene (or Pachynema)

In pachytene, synapsis is complete and the paired chromosomes appear as thicker threads with clearly visible chromomeres (Fig. IX.2c). The

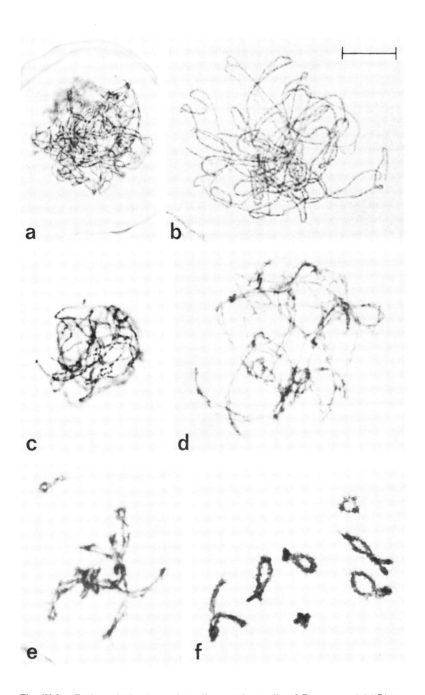

Fig. IX.2. Early meiotic stages in pollen mother cells of *Eremurus*. (a) "Distance pairing" stage; (b) zygotene; (c) pachytene; (d) early diplotene; (e) second contraction in diplotene; (f) late diplotene (Feulgen squash; bar = 15μ) (Therman and Sarto, 1977).

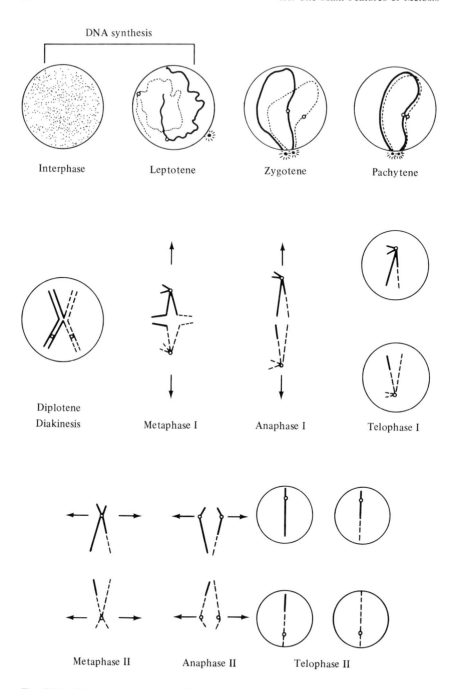

Fig. IX.3. Diagram of meiosis with one pair of chromosomes forming a single chiasma. Note: the chromosomes are *double* after the DNA synthesis although they have been drawn single in zygotene and pachytene.

Fig. IX.4. Zygotene in three pollen mother cells of *Eremurus*; paired and unpaired segments clearly visible (Feulgen squash; bar = 10μ; Therman and Sarto, 1977).

two paired homologues form a *bivalent*. The length of a bivalent at pachytene is estimated to be about one-fifth the length of the same chromosome in leptotene. There is evidence that the chromomere pattern corresponds to the G-bands in the same chromosomes during mitosis. In many organisms the bouquet organization is still visible, whereas in others (most plants) it has disintegrated. Naturally the bouquet is much easier to detect if it persists into pachytene. Crossing-over, which consists of an exchange of homologous segments between two of the four chromatids, takes place in pachytene, although the results of this process cannot be seen until the next meiotic stage, diplotene.

Diplotene (or Diplonema)

The shortening and thickening of chromosomes, which take place between leptotene and pachytene, continue in diplotene. The two homologous chromosomes forming a bivalent begin to repel each other until they are held together only at the points of crossing-over, or as these are called in cytological terms, *chiasmata* (singular: *chiasma*) (Fig. IX.2d-f).

A special type of diplotene chromosome, the lampbrush chromosome, has been studied, especially in the giant oocytes of amphibians. The bivalents are enormously extended, and from each chromomere a pair of DNA loops extends, giving the chromosomes their "lampbrush" appearance. An intensive synthetic activity goes on in the loops. The RNA presumably synthesized under the contol of the DNA in the loops is extruded into the nucleoplasm. At specific loops, DNA rings—apparently copies of the loops—are synthesized and extruded into the nucleoplasm. Here the rings form additional nucleoli, that, in the amphibian oocytes, may number from 200 to 300. It is probable that a stage corresponding to the lampbrush diplotene occurs in the meiocytes of many other organisms. Indications of this phenomenon are seen in the "hairy" appearance of diplotene chromosomes in both plants and animals. This is especially clear in many grasshoppers.

At some point in mid-diplotene, most of the bivalents collect into one group in the middle of the nucleus (Fig. IX.2e). The significance of this stage (the second contraction), is not known.

Diakinesis

Chromosome condensation reaches its final stage in diakinesis. The hairy appearance characteristic of diplotene disappears, and the bivalents are smooth and compact. They seem to be attached to the nuclear membrane, whereas until this stage they appeared to lie free inside the nucleus.

Metaphase I

The nuclear membrane vanishes, and during a short prometaphase stage the bivalents collect into a metaphase plate at the midpoint of the spindle, which has now been formed between the two centrioles (Fig. IX.5a and II.2b). The two centromeres of each bivalent orient themselves toward different poles. The maternal and paternal centromeres are oriented at random, forming the material basis for the independent mendelian segregation of genes.

The shortening and thickening of chromosomes between leptotene and metaphase I are largely caused by the spiralization of the chromosome strand. Compared with mitotic chromosomes, there is an additional supercoil that is beautifully visible in large plant chromosomes, such as those found in *Scilla* or *Tradescantia*.

Anaphase I

In the first meiotic anaphase, the undivided centromeres with two chromatids attached to each, move to opposite poles.

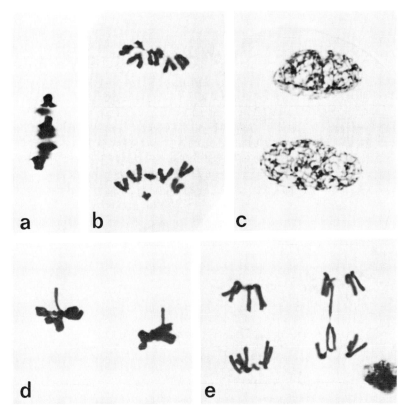

Fig. IX.5. Late meiotic stages in pollen mother cells of *Eremurus*. (a) Sideview of metaphase I; (b) anaphase I; (c) interkinesis; (d) sideview of metaphase II; (e) II anaphase (Feulgen squash).

Telophase I

In the first meiotic telophase, each of the two chromosome groups contains the haploid number of centromeres. If crossing-over has occurred, the two chromatids of a chromosome are not identical. Therefore these chromosomes are different from their mitotic counterparts.

Interkinesis

In many organisms including *Eremurus,* the first meiotic telophase is followed by a short interphase stage, the *interkinesis* (Fig. IX.5c). However, during this stage the chromosomes do not replicate and thus have the same basic structure in prophase as they had in the preceding telophase. In those organisms (grasshoppers, for example) that have no interkinesis, the chromosome groups of anaphase I become the metaphase plates for the second meiotic division.

Meiotic Division II

In organisms both with and without interkinesis, the second meiotic division is indistinguishable from an ordinary, although haploid, mitosis (Fig. IX.5d and e). However, the chromosomes in metaphase II often differ from the mitotic chromosomes in the same organism by being shorter and having widely separated chromatids held together only at the centromere. In anaphase II the centromeres divide and move to the poles with one chromatid attached to each daughter centromere. The end result is four nuclei each with the haploid chromosome complement of which, however, no two are identical in gene content.

In some organisms, so-called terminalization of chiasmata occurs between diplotene and metaphase I. Chiasmata move toward the chromosome ends and apparently decrease in their number by slipping off the ends. This is a different phenomenon from the sliding of chiasmata toward the chromosome ends that regularly takes place in anaphase I. Because true terminalization of chiasmata, which occurs between diplotene and metaphase I, is probably a much rarer phenomenon than has been assumed (cf. Hultén et al, 1978), many of the older claims should be reinvestigated. As far as we know, terminalization does not have any biological meaning

After the premeiotic DNA synthesis, the amount of DNA in the nucleus is the same as in the mitotic prophase, 4C. In anaphase I it is reduced to 2C. Since a DNA synthesis never takes place in interkinesis, the DNA amount in each of the four nuclei resulting from meiosis is 1C, which is characteristic of the gametes.

Extensive biochemical studies of meiosis have been made, especially by the group of H. Stern (cf. Stern and Hotta, 1974). One of the many features distinguishing meiosis from mitosis is that 99.7 percent of the DNA is synthesized in premeiotic interphase or early leptotene. The remaining 0.3 percent of the DNA synthesis occurs during zygotene and seems to play an important role in chromosome pairing. A small amount of repair DNA synthesis occurs in pachytene in connection with crossing-over (cf. Hotta et al, 1977).

The duration of the meiotic stages differs markedly from the mitotic ones in the same organism. Thus the premeiotic DNA synthesis lasts about twice as long as the mitotic S period. Pachytene lasts eight days in the male mouse and 16 days in man, whereas the duration of the mitotic prophase can be measured in hours. Diplotene in the oocytes is a special case; in amphibians the lampbrush stage lasts about six months; in human oocytes it may last as long as 45 years (to be described later).

References

Darlington CD, La Cour LF (1976) The handling of chromosomes, 6th edn. Wiley, New York

Hotta Y, Chandley AC, Stern H (1977) Meiotic crossing-over in lily and mouse. Nature 269: 240–242

Hultén M, Luciani JM, Kirton V, et al (1978) The use and limitations of chiasma scoring with reference to human genetic mapping. Cytogenet Cell Genet 22: 37–58

John B (1976) Myths and mechanisms of meiosis. Chromosoma 54: 295–325

Oksala T, Therman E (1958) The polarized stages in the meiosis of liliaceous plants. Chromosoma 9: 505–513

Stern H, Hotta Y (1974) Biochemical controls of meiosis. Ann Rev Genet 7: 37–66

Therman E, Sarto GE (1977) Premeiotic and early meiotic stages in the pollen mother cells of *Eremurus* and in human embryonic oocytes. Hum Genet 35: 137–151

Walters MS (1970) Evidence on the time of chromosome pairing from preleptotene spiral stage in *Lilium longiflorum* "Croft." Chromosoma 29: 375–418

X
Details of Meiosis

Structure of Chiasmata

A diplotene bivalent with one chiasma, as seen under the microscope, can be interpreted according to either the partial chiasma-type theory or the two-plane theory (Fig. X.1) (cf. Darlington, 1937). The latter assumes that the paired chromatids on the distal side of a chiasma belong to different chromosomes, one paternal and one maternal. According to the chiasma-type theory, two maternal chromatids and two paternal chromatids remain with each other on both sides of the chiasma; in other words, a chiasma is the result of crossing-over. Observations of normal bivalents do not allow one to distinguish between these two possibilities.

Heteromorphic bivalents provide one of the means of distinguishing between the two theories. If one of the homologues has an added segment resulting from a translocation, such a heteromorphic bivalent would have the configuration shown in Fig. X.1 (top), if the two-plane theory is correct. However, such a configuration has never been observed. Instead heteromorphic bivalents always have the configuration shown in Fig. X.1 (bottom), which would be expected based on the chiasma-type theory.

Number of Chiasmata

The typical bivalent has at least one chiasma. If an organism has chromosomes of different sizes, the larger ones usually show a higher number of chiasmata, sometimes as many as six or eight. At metaphase I, a bivalent with one chiasma is either cross-shaped, or a more-or-less

Two-plane-theory:
chiasma may
cause crossing-over

Critical test:
expected appearance of a
heteromorphic bivalent

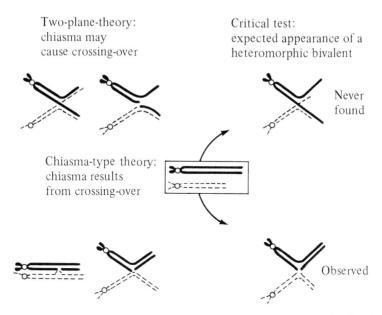

Never
found

Chiasma-type theory:
chiasma results
from crossing-over

Observed

Fig. X.1. Two-plane versus chiasma-type theory as an explanation for the relationship of chiasma and crossing-over.

terminal chiasma connects the homologues end to end. Two chiasmata often result in a ring-shaped bivalent. In a bivalent with several chiasmata, the loops between them are at right angles to each other so that such a configuration resembles a chain.

We still do not know why bivalents (and especially multivalents in polyploids) do not interlock in cases in which pairing has started at several points and more than one chiasma is present. In fact, interlocking bivalents are exceedingly rare. It is possible that the "distance pairing" stage or the bouquet formation, or both represent some type of homologue sorting process before synapsis takes place; this might prevent interlocking when the homologues pair (cf. Oksala and Therman, 1958). Electron-microscopic studies also indicate that the homologous chromosomes align themselves in some way before pairing actually takes place (Westergaard and von Wettstein, 1972).

Although the first chiasma is presumably formed at a random point, the next one cannot arise in the immediate vicinity but has to be a certain distance away from the first chiasma. This phenomenon is called *chiasma interference*. That crossing-over, which is determined by genetic means, shows a similar interference, supplies one of the proofs that chiasmata are the cytological consequence of crossing-over. The correspondence of chiasmata and cross-overs has been demonstrated by autoradiography in the grasshopper (Jones, 1971). Furthermore, in corn the total number of cross-overs (as determined by genetic means) was one-half the number of chiasmata, just as expected (Whitehouse, 1969).

Chiasmata may also be more-or-less strictly localized in a certain chromosome region. Thus in some species of the liliaceous genera *Paris, Allium,* and *Fritillaria* as well as in certain grasshoppers, each bivalent has only one chiasma, which is localized near the centromere. In other plants, as well as in some newts, the chiasmata are situated at the ends of chromosomes (cf. Darlington, 1937; John and Lewis, 1965; White 1973).

In exceptional cases, crossing-over can "go wrong," leading to a dicentric and an acentric chromatid. Such configurations are called U-type exchanges, because they resemble the letter U (cf. Rees and Jones, 1977).

Synaptonemal Complex

The *synaptonemal complexes,* which can be studied with the electron microscope, are essential elements in chromosome pairing and crossing-over. The basic units of the synaptonemal complex are the lateral elements. These structures consist of protein and RNA and arise in the nucleoplasm during the leptotene stage. When chromosome pairing starts, one end of each homologue becomes attached to the nuclear membrane and the two homologues become roughly aligned, a lateral element attaches to each of them. The lateral elements pair to form the synaptonemal complex, which includes a central element that completes the synapsis (Fig. X.2). The formation of a synaptonemal complex between a pair of homologues is initiated at several points. During diplotene the synaptonemal complexes separate from the bivalents and either collect into so-called polycomplexes or disintegrate (cf. Wester- gaard and von Wettstein, 1972; Luykx, 1974).

Synaptonemal complexes have been found in all organisms in which chromosome pairing and four-strand crossing-over occur (Westergaard and von Wettstein, 1972). They obviously play an important role in the effective chromosome pairing that is a prerequisite for crossing-over. For example, they are absent in the meiocytes of the *Drosophila* male (an organism in which crossing-over does not take place), but they are present in the female of the same species (in which chiasmata are formed). Although synaptonemal complexes are apparently needed for crossing-over, their presence does not guarantee the occurrence of this process. They have been found in the meiocytes of haploid plants where no chiasmata are formed (cf. Gillies, 1975). Furthermore, a short syn- aptonemal complex is formed between the human X and Y chromosomes although no evidence exists for crossing-over between them.

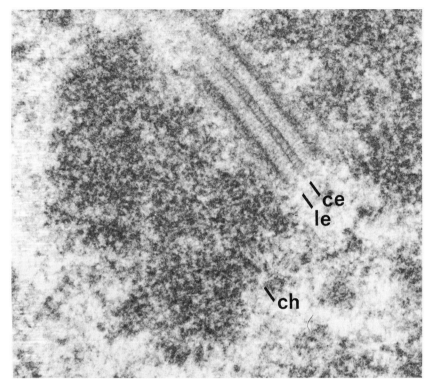

Fig. X.2. Synaptonemal complex from the spermatocyte of the spittle bug *Philaemus* (le = lateral element, ce = central element, ch = chromatin) (courtesy of H Ris).

Although the discovery of the ubiquity of the synaptonemal complex is important for the interpretation of meiotic phenomena, numerous problems presented by chromosome pairing and crossing-over remain unresolved. The chromosomes are hundreds of times longer than their synaptonemal complexes and it is unclear how the exact pairing of the corresponding DNA sequences and subsequent crossing-over take place between them. It is assumed that segments of DNA are trapped in the central region of the synaptonemal complex, and that the chiasmata are formed between these.

Meiotic Behavior of More Than Two Homologous Chromosomes

Meiosis in polyploids has been analyzed extensively, especially in plants. It is of little significance in human cytogenetics, since the only polyploids known—triploids and tetraploids—as a rule die prenatally and survive birth only exceptionally. However, the early meiotic stages in oocytes of a triploid human fetus have been analyzed by Luciani et al (1978).

Women with Down's sydrome (21 trisomy) are the only individuals with autosomal trisomy known to have reproduced; their offspring consist of children with Down's syndrome and normal children in a ratio that does not differ significantly from the expected 1:1 (cf. Hamerton, 1971, p. 214).

Three homologous chromosomes usually pair to form a *trivalent* in meiosis (Fig. X.3). However, at any one point only two of them synapse (cf. Darlington, 1937). As shown in Fig. X.3, two chromosomes (the thick and the interrupted line) start to pair, then there is an exchange of partners (the thick pairs with the thin), and at the second exchange the first two chromosomes (the thick and the interrupted line) resume pairing. If chiasmata are formed in the three paired segments, the diplotene configuration shown in Fig. X.3 results. An alternate centromere orientation gives rise to the metaphase I trivalent in Fig. X.3. Depending on

Fig. X.3. Pairing of three homologous chromosomes. Note: the chromosomes are *double* in pachytene although they are drawn single.

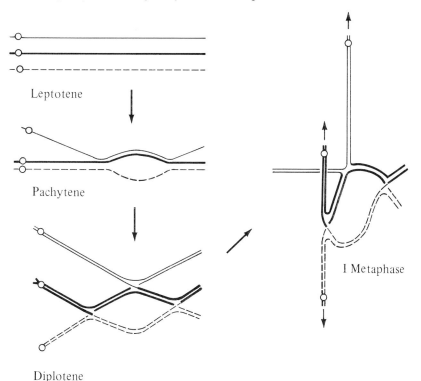

Leptotene

Pachytene

Diplotene

I Metaphase

the number and length of the paired segments and the number and location of chiasmata, trivalents come in a variety of shapes. Three homologues may also form a bivalent and a univalent. Of the three homologues usually two go to one pole and one to the other in anaphase I. This type of segregation is called *secondary nondisjunction*.

Four homologous chromosomes may form a *quadrivalent,* or two bivalents, or a trivalent and a univalent. Even more possibilities exist in higher polyploids, such as hexaploids, in which chains or rings of six chromosomes are observed. Because meiotic segregation in polyploids is often irregular, the gametes may receive inviable chromosome constitutions, resulting in partial sterility.

Human Meiosis

In the human male, as in the males of most plants and animals, the four meiotic products (spermatids), form four functional gametes. Meiosis in the oocyte results in one functional egg cell and three very small cells. In animals these are called polar bodies and degenerate, whereas in plants the three nuclei are included in the structure of the embryo sac.

From a cytologist's point of view, meiosis in human oocytes and spermatocytes complement each other in that the meiotic prophase stages have been studied—and are much clearer—in the female, whereas our information on stages from diplotene onward comes mainly from the male.

Premeiotic and Early Meiotic Stages in Man

The early meiotic stages in the human female are found in 12- to 16-week-old embryos. The last premeiotic mitosis, in which the metaphase chromosomes appear short and thick, is followed by an interphase in which most or all of each chromosome remains condensed as a chromocenter. These chromocenters unravel to form the leptotene threads. The meiotic DNA synthesis occurs in early leptotene, as in *Eremurus* pollen mother cells. Between leptotene and zygotene there is a short "distance pairing" stage (Therman and Sarto, 1977)—again as in *Eremurus*—during which the homologues appear to lie side by side but at a distance (Fig. X.4a).

The diplotene stage in human oocytes is a very long phase, lasting from the embryonic stage until the egg cell is released in the adult female. This type of diplotene, called the dictyotene, represents a period of enormous growth of the cell. The chromosomes are greatly extended and invisible with most cytological techniques. Dictyotene corresponds to

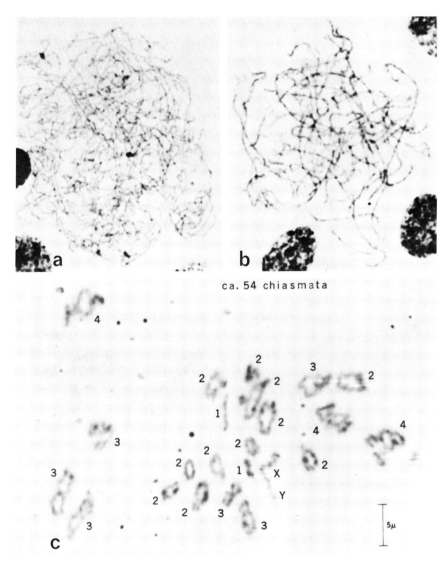

ca. 54 chiasmata

Fig. X.4. a-c. Meiotic stages in human oocytes. (a) "Distance pairing" stage; (b) zygotene (Therman and Sarto, 1977); (c) diakinesis from a human spermatocyte (Feulgen squash).

the lampbrush stage and in addition to the normal nucleoli, 15–20 extra nucleoli are formed, which seem to be attached to the heterochromatic centric regions of the chromosomes, especially that of chromosome 9 (Stahl et al, 1975). A similar diffuse diplotene stage is also described in the spermatocytes of insects and many other organisms.

The bivalent consisting of the two X chromosomes behaves in the oocytes in the same way as the autosomal bivalents do, that is, without showing any heteropycnotic condensation. In this respect the X chromosomes behave differently in the oocytes than in female somatic cells where one of them acts heterochromatically. In the spermatocytes of the male, both the X and the Y chromosomes are heteropycnotically condensed in the meiotic prophase from zygotene to diplotene.

Human Meiotic Stages from Diplotene on

As mentioned earlier, the later meiotic stages in man have been studied almost exclusively in the male. The mean chiasma frequency in human spermatocytes, as determined from diakinesis and metaphase I stages, is 50.61 (SD 3.87) (Hultén, 1974). The number of chiasmata in individual cells varies from 43 to 60. As seen in Fig. X.4c, the largest human bivalents usually have four chiasmata; the medium-sized, two to three; and the smallest ones, one each. In many organisms the chiasma frequency in the oocytes and the spermatocytes is significantly different. Whether this applies to man too is not known, since so little information exists about the later meiotic stages in the female.

The second meiotic division in the spermatocytes follows the usual scheme, as described in the pollen mother cells of *Eremurus*.

Behavior of X and Y Chromosomes

The X and Y chromosomes in man, as in most other animals, appear heteropycnotic from zygotene to diplotene. In pachytene, in which a clear bouquet organization is visible, they form a paired clump called the XY body or (less appropriately) the "sex vesicle" (cf. Solari, 1974). In diplotene, the sex chromosomes cease to show a heterochromatic behavior and during diplotene-metaphase I, the XY body appears similar to the autosomal bivalents except for its asymmetrical structure.

The X and Y chromosomes of both man and mouse are paired end to end, short arm to short arm in man, and long arm to long arm in mouse. The pairing behavior of the two sex chromosomes shows wide variation in mammals: the X and the Y may have a considerable segment in common where pairing and crossing-over take place as in the autosomes. In addition, they have a differential segment in which the sex determinants are situated. A real chiasma between the sex chromosomes has been observed, for instance, in several hamster species representing different genera (cf. Solari, 1974). In both mouse and man the pairing segment is small and although a short synaptonemal complex is formed, it is not likely that crossing-over takes place. An extreme case is the common field vole, *Microtus agrestis,* in which the X and the Y do not pair at all but still show regular segregation (Zenzes and Wolf, 1971). The earlier claims of crossing-over between the X and Y chromosomes in man, leading to partial sex-linkage of some genes, have been dropped for lack of evidence.

In the first meiotic anaphase, the X and the Y chromosomes segregate regularly and in the second division they divide mitotically, which leads to the formation of two spermatozoa with an X chromosome and two with a Y.

References

Darlington CD (1937) Recent advances in cytology, 2nd edn. Churchill, London

Gillies CB (1975) Synaptonemal complex and chromosome structure. Ann Rev Genet 9: 91–109

Hamerton JL (1971) Human cytogenetics, Vol II. Academic, New York

Hultén M (1974) Chiasma distribution at diakinesis in the normal human male. Hereditas 76: 55–78

John B, Lewis KR (1965) The meiotic system. In: Protoplasmatologia, Vol VI, Fl. Springer, New York

Jones GH (1971) The analysis of exchanges in tritium-labelled meiotic chromosomes. Chromosoma 34: 367–382

Luciani JM, Devictor M, Boué J, et al (1978) The meiotic behavior of triploidy in a human 69,XXX fetus. Cytogenet Cell Genet 20: 226–231

Luykx P (1974) The organization of meiotic chromosomes. In: Busch H (ed) The cell nucleus, Vol II. Academic, New York pp 163–207

Oksala T, Therman E (1958) The polarized stages in the meiosis of liliaceous plants. Chromosoma 9: 505–513

Rees H, Jones RN (1977) Chromosome genetics. University Park Press, Baltimore

Solari AJ (1974) The behavior of the XY pair in mammals. In: Bourne GH, Danielli JF (eds) International review of cytology, Vol 38. Academic, New York, pp 273–317

Stahl A, Luciani JM, Devictor M, et al (1975) Constitutive heterochromatin and micronucleoli in the human oocyte at the diplotene stage. Humangenetik 26: 315–327

Therman E, Sarto GE (1977) Premeiotic and early meiotic stages in the pollen mother cells of *Eremurus* and in human embryonic oocytes. Hum Genet 35: 137–151

Westergaard M, von Wettstein D (1972) The synaptinemal complex. Ann Rev Genet 6: 71–110

White MJD (1973) Animal cytology and evolution, 3rd edn. University Press, Cambridge, England

Whitehouse HLK (1969) Towards an understanding of the mechanism of heredity. Arnold, London

Zenzes MT, Wolf U (1971) Paarungsverhalten der Geschlechtschromosomen in der männlichen Meiose von *Microtus agrestis*. Chromosoma 33: 41–47

XI
Meiotic Abnormalities

Nondisjunction of Autosomes

Meiosis is a much more complicated chain of events than mitosis. Consequently the chances of failure during some part of the process are also more numerous. Most meiotic irregularities lead to nondisjunction or to chromosome loss, which in turn results in the formation of gametes with aneuploid chromosome numbers. Nondisjunction refers to any process resulting in two chromosomes, which ought to segregate to opposite poles, going to the same pole.

Some of the possible meiotic aberrations are shown in Fig. XI.1, which also shows the fate of two allelic genes, A and a. In the first case, the two chromosomes of a bivalent have failed to disjoin and therefore go to the same pole. In the second meiotic division, the two homologues divide normally resulting in the formation of two disomic and two nullosomic gametes. In the following example, the chromosomes have paired normally but no chiasma has formed. Consequently the paired homologues fall apart and appear as univalents.

Univalents may behave in different ways; they may drift at random to the two poles in the first division and divide regularly in the second. Alternatively, they may divide mitotically in anaphase I and, being single chromatids in the second division, they cannot divide any more and so drift at random to the poles or misdivide. Similar behavior is also exhibited by univalents if they result from an original failure to pair in zygotene. One possibility is that the drifting univalents do not reach the poles but remain as laggards and are lost.

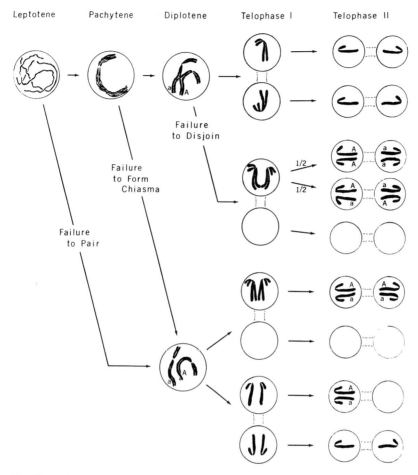

Fig. XI.1. Diagram of the processes resulting in meiotic nondisjunction (see text) (Patau, 1963).

Nondisjunction of Sex Chromosomes

Since the XX bivalent in mammalian oocytes behaves as the autosomes do, the two X chromosomes in the female exhibit the same irregularities in pairing and segregation as the rest of the chromosomes.

The behavior of the X and the Y chromosomes in the male, on the other hand, differs from the autosomes. Although a short synaptonemal complex is formed between the two sex chromosomes, their end-to-end attachment seems to be different in kind from a typical terminal chiasma. This is confirmed by the fact that the X and the Y are found much more often as univalents than are the small autosomes with one terminal chiasma.

In a study of meiosis in 53 normal men, the mean fraction of spermatocytes in which the X and the Y were unpaired was 3.2 percent (McDermott, 1971). Much higher frequencies of separation of the sex chromosomes have also been reported (cf. Solari, 1974, p. 304), but many of these earlier claims sound unrealistic. In normal mice, 8 percent to 10 percent of the spermatocytes are observed to have a separate X and Y (cf. Rapp and Therman, 1977). An interesting although so far unexplained observation is that the pairing of the two sex chromosomes seems to be a prerequisite for the completion of meiosis and, therefore, for the formation of functional spermatozoa in man as well as in the mouse.

Misdivision of the Centromere

Chromosomes remaining as univalents in the first meiotic division may undergo so-called *misdivision*. Univalents may orient wrongly on the spindle, or their abnormal condition during the first meiotic division may itself lead to an abnormal division. Whereas in normal mitosis the centromere divides to separate the two sister chromatids, in misdivision it divides cross-wise so that one whole chromosome arm goes to one pole and the other arm to the opposite pole (Fig. XI.2). The human X chromosome may be especially liable to misdivision, which results in the two long-arm chromatids separating from the two short-arm chromatids. However, of the two products, only the long-arm isochromosome, i(Xq), seems to be viable (cf. Therman and Patau, 1974).

Misdivision and its consequences have been studied especially in plants (cf. Darlington, 1939; Sears, 1952). In addition to the centromere dividing cross-wise, it can separate into four parts, which will give rise to two long-arm telocentric chromosomes and two short-arm telocentrics. Again, in man, only the long-arm telocentric of the X chromosome has been found; the short-arm telocentric has never been observed. Misdivision of centromeres may also take place in the second meiotic division or in mitosis, although this is much rarer than its occurrence in the first meiotic division.

Environmental Causes of Meiotic Nondisjunction

The causes of nondisjunction may be environmental or genetic. Experiments conducted on plants and animals reveal that a variety of environmental agents, both physical and chemical, affect meiosis. For example,

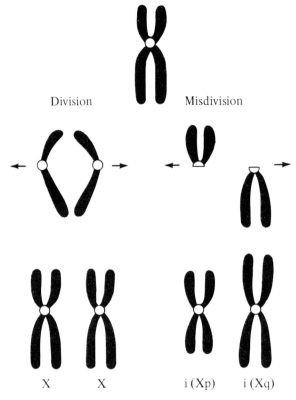

Fig. XI.2. Misdivision of the human X chromosome resulting in the formation of a long-arm isochromosome i(Xq) and a short-arm isochromosome i(Xp) (presumably inviable).

heat shocks prevent chromosome pairing by causing the chromosomes to remain as univalents, which leads to a totally irregular meiosis. The same phenomenon is caused by cold treatment in mouse embryos (Karp and Smith, 1975). Similarly, ionizing radiation increases nondisjunction in *Drosophila* (cf. Uchida, 1977).

A manifold of substances are known to cause chromosome breakage (cf. Kihlman, 1966; Rieger and Michaelis, 1967). Although less attention has been paid to nondisjunction, most of the chromosome-breaking substances probably also give rise to occasional abnormal segregation in mitosis or meiosis. For instance, alkylating leukemia drugs increase meiotic nondisjunction in mice (Röhrborn, 1971; Vogel et al, 1971).

Another class of substances, of which the most widely used is the alkaloid colchicine, specifically destroys the spindle structure. Instead of collecting into a metaphase plate, the chromosomes are scattered around the cell, which in milder cases leads to nondisjunction and in extreme cases to restitution, giving rise to a tetraploid cell. As early as 1939, Levan showed that colchicine interfered with chromosome pairing in the meiosis of the onion. Recently Shepard et al (1974) established that colchicine largely destroys the lateral elements of the synaptonemal complex. As a result, only limited stretches of chromosomes pair effectively and form chiasmata. The cytological effects of colchicine should be taken into consideration when this alkaloid is used as a gout medicine for patients of reproductive age.

We have much less definitive information about the causes of meiotic nondisjunction in man than in plants and animals. However, the results of this process—trisomic and monosomic offspring and chromosomally abnormal spontaneous abortions—demonstrate its existence. It is not far-fetched to assume that the same agents that cause nondisjunction in other organisms also affect human meiosis.

Maternal Age

A well established cause of nondisjunction in man is increased maternal age, which, as far as the gametes are concerned, can be classified as an environmental influence. At the maternal age of 20 years, the incidence of 21-trisomic children is 0.4 per 1000 newborns; for women age 45 years and over, the risk increases to 17 per 1000 newborns. A similar maternal age effect is found for 18- and 13-trisomic children as well as for offspring with XXX and XXY sex chromosome constitutions. However, this rule does not apply to Turner's syndrome with the 45,X chromosome constitution. Spontaneous abortions with aneuploid chromosome numbers also occur more frequently in older women (cf. Carr, 1971). It is estimated that if women over 35 years of age refrained from having offspring, the incidence of children with abnormal chromosome numbers would decrease by one-third to one-half.

However, with chromosome banding techniques and more refined statistical methods it is possible to demonstrate that the earlier belief that the father's age played no role in the birth of aneuploid children is simply untrue. Identification of the individual 21 chromosomes with banding techniques shows that nondisjunction for this chromosome also occurs in the father: possibly as many as one-third of 21-trisomic children owe their origin to paternal nondisjunction (Wagenbichler et al, 1976; Uchida, 1977; Mattei et al, 1979), and from age 55 onward their frequency increases with paternal age (Stene et al, 1977).

In mouse oocytes the frequency of chiasmata is decreased and the incidence of univalents increased with increasing age of the female (Henderson and Edwards, 1968). These results were confirmed by Luthardt et al (1973) who also observed that the univalent formation was nonrandom, involving mainly the smaller chromosomes. This finding agrees with observations on many organisms that the smallest chromosomes of the complement with at most one chiasma are most likely to remain as univalents.

The mechanism by which the age of the female acts on meiosis is not known. Crossing-over takes place during pachytene in embryonic oocytes, and it seems highly improbable that the number of chiasmata would decrease even if the diplotene stage lasts for a very long time. A more plausible explanation is that there is a gradient in the ovary, and the egg cells that are released earlier have more chiasmata than those released later.

Another factor that increases nondisjunction in women is preconception exposure to ionizing radiation (Uchida, 1977). Although in a couple of studies no such effect could be established, the accumulated evidence from several others speaks strongly for the assumption that radiation must be taken seriously as a possible cause for the birth of human trisomics. Radiation seems to increase nondisjunction, especially in older women and in older female mice (Uchida, 1977).

Genetic Causes of Nondisjunction

In many organisms, for instance in *Drosophila* and maize, genes are known that upset the normal course of meiosis by affecting either synapsis or chiasma formation. In man, a number of abnormal individuals are described who show aneuploidy for more than one chromosome. This fact naturally reflects a more serious disturbance of meiosis than simple nondisjunction. For example, both an extra 21 and an extra 18 were found in an extremely unfortunate infant who died 8 h after birth (Gagnon et al, 1961). Other combinations are trisomy 21 and an XXX or XXY sex chromosome constitution, or an extra 21 combined with a missing X chromosome (cf. Mikkelsen, 1971). Other types of double aneuploidy are occasionally found.

The occurrence of more than one child with an aneuploid chromosome constitution in the same sibship has led to the conclusion that such families may have an inherent tendency to nondisjunction. Thus the second D-trisomic child ever described had a sister with a 45,X chromosome constitution (Therman et al, 1961). However, the most convincing evidence that some families have a genetic disposition to nondisjunction is from Hecht et al (1964). The study started with 60 families with either an 18-trisomic or D-trisomic offspring. Three of the 18-trisomic children had siblings with 21 trisomy as compared with an expectation of 0.15 mongoloid children for the same group—a highly significant statistical difference. In addition, one of the children with D-trisomy had an uncle with 21 trisomy.

Origin of Diploid Gametes

Tetraploid and triploid abortions, and even a few liveborn children, demonstrate that unreduced gametes are formed and are occasionally able to function. Further, spectrophotometric measurements show that some sperms contain double the usual amount of DNA. Such gametes owe their origin mainly to two processes. The same agents causing nondisjunction may disturb meiosis to the extent that a restitution nucleus, which contains the unreduced diploid complement, is formed in the first meiotic division, or endoreduplication in a gonial cell may lead to one or more tetraploid meiocytes. If such a cell subsequently under-goes normal meiosis, it may give rise to diploid gametes.

By using banding techniques to extract informative data on 10 triploid abortuses, Kajii and Niikawa (1977) found that: one originated in mater-nal first meiotic division; five apparently resulted from two sperms fertilizing the same egg; two owed their origin to an aberration in either the second paternal division or the first mitotic division; and the last two were of undefined paternal origin. One tetraploid abortion obviously came about by the suppression of the first cleavage division. Similar results were obtained by Jacobs et al (1978) in a study on 26 triploid abortions. To quote their paper (p. 56): "The best fit for the data using a maximum-likelihood method was that 66.4% of the triploids were the result of dispermy, 23.6% the result of fertilization of a haploid ovum by a diploid sperm formed by failure of the first meiotic division in the male and 10% the result of a diploid egg formed by failure of the first maternal meiotic division."

References

Carr DH (1971) Chromosomes and abortion. In: Harris H, Hirschhorn K (eds) Advances in human genetics. Plenum, New York

Darlington CD (1939) Misdivision and the genetics of the centromere. J Genet 37: 341–364

Gagnon J, Katyk-Longtin N, de Groot JA, et al (1961) Double trisomie autoso-mique à 48 chromosomes (21+18). L'Union Méd Canada 90: 1–7

Hecht F, Bryant JS, Gruber D, et al (1964) The nonrandomness of chromosomal abnormalities. New Eng J Med 271: 1081–1086

Henderson SA, Edwards RG (1968) Chiasma frequency and maternal age in mammals. Nature 218: 22–28

Jacobs PA, Angell RR, Buchanan IM, et al (1978) The origin of human triploids. Ann Hum Genet 42: 49–57

Kajii T, Niikawa N (1977) Origin of triploidy and tetraploidy in man: 11 cases with chromosome markers. Cytogenet Cell Genet 18: 109–125

Karp LE, Smith WD (1975) Experimental production of aneuploidy in mouse oocytes. Gynecol Invest 6: 337–341

Kihlman BA (1966) Actions of chemicals on dividing cells. Prentice-Hall, Englewood Cliffs, New Jersey

Levan A (1939) The effect of colchicine on meiosis in *Allium*. Hereditas 25: 9–26

Luthardt FW, Palmer CG, Yu P-L (1973) Chiasma and univalent frequencies in aging female mice. Cytogenet Cell Genet 12: 68–79

Mattei JF, Mattei MG, Ayme S, et al (1979) Origin of the extra chromosome in trisomy 21. Hum Genet 46: 107–110

McDermott A (1971) Human male meiosis. Can J Genet Cytol 13: 536–549

Mikkelsen M (1971) Down's syndrome: current stage of cytogenetic research. Humangenetik 12: 1–28

Patau K (1963) The origin of chromosomal abnormalities. Pathol Biol 11: 1163–1170

Rapp M, Therman E, Denniston C (1977) Nonpairing of the X and Y chromosomes in the spermatocytes of BDF_1 mice. Cytogenet Cell Genet 19: 85–93

Rieger R, Michaelis A (1967) Die Chromosomenmutationen. Gustav Fischer, Jena

Röhrborn G (1971) Chromosome aberrations in oogenesis and embryogenesis of mammals and man. Arch Toxikol 28: 115–119

Sears ER (1952) Misdivision of univalents in common wheat. Chromosoma 4: 535–550

Shepard J, Boothroyd ER, Stern H (1974) The effect of colchicine on synapsis and chiasma formation in microsporocytes of *Lilium*. Chromosoma 44: 423–437

Solari AJ (1974) The behavior of the XY pair in mammals. In: Bourne GH, Danielli JF (eds) International review of cytology, Vol 38: Academic, New York, pp 273–317

Stene J, Fischer G, Stene E (1977) Paternal age effect in Down's syndrome. Ann Hum Genet 40: 299–306

Therman E, Patau K (1974) Abnormal X chromosomes in man: origin, behavior and effects. Humangenetik 25: 1–16

Therman E, Patau K, Smith DW, et al (1961) The D trisomy syndrome and XO gonadal dysgenesis in two sisters. Am J Hum Genet 13: 193–204

Uchida IA (1977) Maternal radiation and trisomy 21. In: Hook EB, Porter IH (eds) Population cytogenetics. Academic, New York pp 285–299

Vogel F, Röhrborn G, Schleiermacher E (1971) Chemisch-induzierte Mutationen bei Säuger und Mensch. Naturwissenschaften 58: 131–141

Wagenbichler P, Killian W, Rett A, et al (1976) Origin of the extra chromosome no. 21 in Down's syndrome. Hum Genet 32: 13–16

XII
Human Sex Chromosomes

The Y Chromosome

The human male has two sex chromosomes, one X and one Y. The Y chromosome is in the same size range as the two smallest pairs (the G group) of autosomes. However, the following features enable one to distinguish the Y, even with ordinary staining techniques, from the G chromosomes: the short arm does not have satellites, a constriction is sometimes visible in the middle of the long arm, the chromatids of the long arm stick together, and the distal end of the long arm often looks fuzzy. During the mitotic cycle, the Y replicates later than do the G chromosomes.

The Q-banding pattern of the Y chromosome is much more spectacular than that of the G-group chromosomes, thanks to the very bright distal segment of the long arm (Fig. XII.1b). Without causing any apparent phenotypic effects, the length of this segment may range from practically zero, which is exceedingly rare, to two or three times its length in the average Y chromosome. Orientals have been reported to have longer Y chromosomes, on the average, than males of Caucasian extraction (Cohen et al, 1966). The length of the brightly fluorescent distal segment is reflected in the size of a bright Y-body that can be seen in interphase nuclei. Even a sperm with a Y chromosome is distinguishable from X-carrying sperms by its bright fluorescent spot. In cells with two Y chromosomes, two Y-bodies can be seen (Fig. XII.1c).

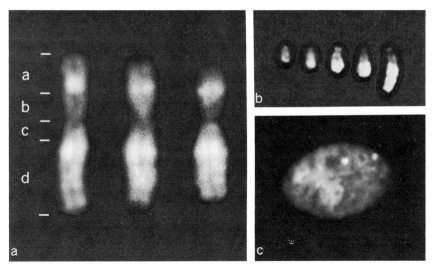

Fig. XII.1. (a) Q-banded human X chromosomes (Therman et al, 1974); (b) variation in the size of the human Y chromosome (Dutrillaux, 1977); (c) two Y bodies in a buccal cell from a 47,XYY male.

Sex Determination in Man

The mechanisms of sex determination may differ greatly, even among closely related groups. A sex chromosome mechanism is only one possibility, and even in it considerable variation occurs. In the plant *Melandrium,* in which abnormal sex chromosome constitutions have been studied extensively, it is the *presence* of the Y chromosome that primarily determines male development. In *Drosophila,* sex is determined by the *ratio* of X chromosomes to the haploid sets of autosomes. Sex determination in man and in other mammals resembles the plant *Melandrium,* rather than the *Drosophila.* A Y chromosome in man guarantees a male phenotype (albeit an abnormal one) even if four X chromosomes are also present. (More than five sex chromosomes appear to be lethal.)

Individuals with an isochromosome Yq in addition to a normal X are females. Phenotypically they are usually identified as belonging to the Turner's syndrome. In other words, they resemble individuals with the 45,X chromosome constitution (for example, Robinson and Buckton, 1971; Magnelli et al, 1974). A girl with several Turner symptoms and a deletion of Yp (she had one X chromosome) was described by Rosenfeld et al (1979). These observations indicate that the male-determining gene, or genes, lie on the short arm of the Y chromosome.

Another type of abnormal Y chromosome, found usually in mosaic individuals with a 45,X cell line, has been described (cf. Cohen et al,

1973). Such chromosomes have two Y long arms, two centromeres, and part of the short arms between them. Individuals with such a chromosome constitution display a phenotype ranging from female to abnormal male, depending on the length of the Yp included and the ratio of the two cell lines.

Dicentric Y chromosomes in which one centromere is inactivated—they resemble an asymmetrical i(Yq)—are found in a few mosaics (Schmid and D'Apuzzo, 1978; Taylor et al, 1978). Very small cross-like sex chromosomes, which are difficult to interpret, have also been described, mostly in males. They may represent Yq− chromosomes, isochromosomes for Yp, or possibly fragments of X chromosomes.

All in all, the evidence is strong for the location of the male-determining genes near the centromere on the Yp, although in some cases the relationship of the structurally abnormal Y chromosome and the phenotype is difficult to interpret.

H-Y Antigen

Recent studies have considerably increased our understanding of the primary (gonadal) sex determination. It has long been clear that the basic plan of an undifferentiated gonad is female. A gene (or genes) on the Y chromosome, probably on its short arm, induces the production of the H-Y antigen, which is a plasma membrane protein (cf. Ohno, 1979). Whether it is this gene that codes for the antigen, or whether the gene has a regulatory function, is not known (cf. Wolf, 1978). The H-Y antigen in turn determines the development of the testis. The differentiated testis tissue secretes testosterone, which brings about the secondary sex differentiation.

Whenever a Y chromosome or testicular tissue is present, the H-Y antigen is also found, and vice versa. Dispersed ovarian cells of rat embryos organize into testicular tissue under the influence of the H-Y antigen (Zenzes et al, 1979), whereas testicular mouse cells stripped of their H-Y antigen sites associate into ovarian structures (Ohno et al, 1978).

A mutant gene in the X-linked Tfm locus causes individuals with an XY sex chromosome constitution to become phenotypic females. This testicular feminization syndrome is discussed later in this chapter (p. 121). The normal allele at the Tfm locus has been assumed to act as a major regulatory gene in secondary sex development (cf. Wachtel, 1979). Whether other, possibly autosomal, genes are involved in this process is not known.

One abnormal group that did not seem to fit into this scheme comprised men with an XX sex chromosome constitution in whom the testicular development is induced by the H-Y antigen. However, at least a partial solution to this riddle is provided by the observation that in about 70 percent of the XX men, a segment, presumably from the Y, is translocated to the short arm of one X chromosome (cf. Breg et al, 1979; Evans et al, 1979). The remaining cases seem to be caused by a Y-autosomal translocation, mosaicism for a Y-containing cell line, or an autosomal gene.

Genes on the Y Chromosome

Apart from the male-determining factor(s) on the Y chromosome, the human Y chromosome seems to be peculiarly "empty," even considering that the Q-bright distal segment consists of constitutive heterochromatin. This is the more striking since holandric (Y-linked) inheritance would be especially easy to notice and to study, and therefore Y-linked genes ought not to escape notice. One of the few genes reported on the Y chromosome causes "hairy ears," which means the growth of hairs on the rim of the earlobe (cf. Dromandraju, 1965). Factors influencing the overall height of the individual and the size of the teeth seem to be located in the long arm of the Y chromosome (Alvesalo, 1978).

Abnormal Y Chromosomes

Apart from the polymorphism of the bright distal region on Yq, a number of structurally abnormal Y chromosomes have been found in man. The isochromosomes for Yq, Yp, and dicentrics in which the two Y chromosomes are attached short arm-to-short arm have already been described. Dicentrics in which the fusion takes place between two long arms also exist. An inversion that places the centromere at the border of the Q-bright region on the Yq does not seem to have an effect on the phenotype or fertility and has been found to be inherited throughout a very big pedigree (Friedrich and Nielsen et al, 1974). Several Y ring chromosomes have also been described.

The X Chromosome

The normal female chromosome complement contains, in addition to 44 autosomes, two X chromosomes that are members of the medium-sized C group. The X chromosome constitutes 5.3 percent of the length of the haploid karyotype, and its centromere index is 0.38. The fluorescent pattern of the X chromosome is shown in Fig. XII.1a. In the short arm, two main regions can be seen: a brighter distal segment (a) and a less Q-bright region (b) next to the centromere. On the other side of the centromere is another Q-dark segment (c), which is shorter than the corresponding region on the Xp (b). The rest of the Xq is Q-bright (d) and is divided into two almost equal parts by a narrow darker band.

X Chromatin or Barr Body Formation

In 1949 Barr and Bertram discovered that the nuclei of the nerve cells of the female cat had a condensed, deeply stained body that was absent in the cells of the males. This X chromatin (also called sex chromatin or Barr body) was later found to represent one X chromosome that is condensed and inactive in the female. The same behavior has been established for one X chromosome in practically all mammalian females. If an individual, female or male, has more than two X chromosomes, all but one of them form Barr bodies.

Fig. XII.2d–f illustrates the behavior of the Barr body-forming X chromosome in interphase and through prophase in human fibroblast cells. Even in prophase it continues to be more condensed than the other chromosomes. Apart from cultured fibroblasts, Barr bodies in man are most often studied in buccal smears (Fig. XII.2b–c), but also in vaginal smears, and sometimes in hair-root follicles. In normal females, 30 percent to 50 percent of the buccal cells show a Barr body, but the counts from different laboratories vary greatly. This relatively low frequency probably results from the fact that many of the cells are dead (the deeper one digs in the buccal mucosa, the higher the incidence). In fibroblasts, the incidence of cells with Barr bodies is more than 90 percent when most of the cells are in the G_2 phase. In polymorphonuclear white blood cells, the inactive X appears as a drumstick-shaped extrusion with the frequency of 1 percent to 10 percent (Fig. XII.2a). Matters pertaining to X chromatin are reviewed by Mittwoch (1974).

The allocycly, or being out of step with the other chromosomes, shown by the inactive X chromosome also expresses itself in other ways. The inactive X replicates late during the S period, as demonstrated by autoradiography. When tritiated thymidine is fed to a cell late in the S phase, the inactive X may be the only labeled chromosome in the next metaphase (Fig. XII.3). In most cells the inactive X is the latest-labeling of the C-group chromosomes.

$\longrightarrow$

Fig. XII.2. Behavior of the inactive X chromosome(s). (a) A drumstick in a polymorphonuclear blood cell; (b) two Barr bodies in the buccal nucleus of a 47,XXX woman; (c) one Barr body in a normal woman; (d) inactive X chromosome in early prophase; (e) Barr body in a fibroblast; (f) inactive X chromosome in late prophase (Feulgen staining, bar = 5 μ).

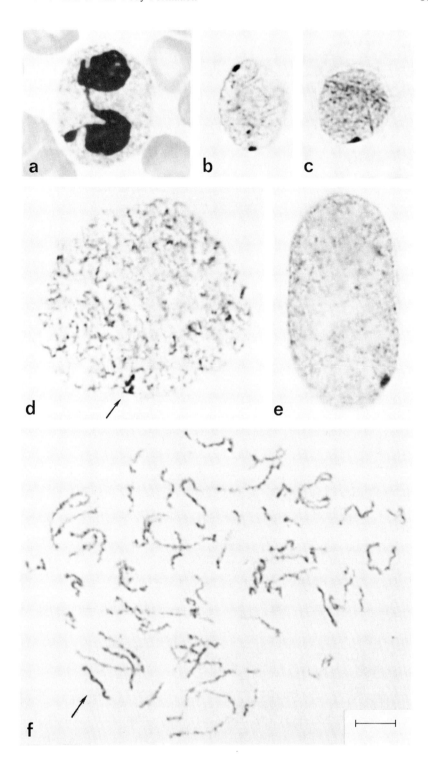

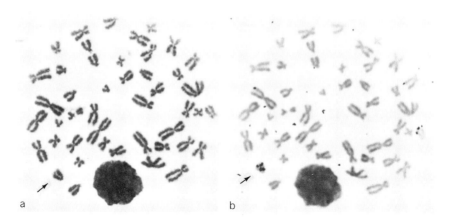

Fig. XII.3. Late-replicating tel(Xq) (arrows) in a woman with 46,X,tel(Xq). (a) Metaphase (orcein staining); (b) same cell after autoradiography; only the abnormal X chromosome is labelled.

A more refined technique to distinguish the Barr body-forming X in metaphase has been invented by Latt (1973). It involves growing the cells for 40–44 h in a medium containing BrdU and substituting thymidine for the last 6–7 h before fixation. The inactive X appears very bright when stained with fluorescent stains Hoechst 33258 or coriphosphine O (Fig. XII.4). Stained with Giemsa techniques it is darker than the other X. The inactive X is often shorter in metaphase than the active one and may show a differential staining even with ordinary cell culture methods (Takagi and Oshimura, 1973; Sarto et al, 1974).

Inactivation of the X Chromosome

Unlike the Y chromosome, the human X seems to contain genes in proportion to its length. Despite the fact that males have one and females two X chromosomes, the sexes do not differ from each other very much, apart from sexual development and secondary sex characteristics. In many other animal groups, gender differences are considerable. This relative similarity of the male and female in mammals in general is achieved by a mechanism of dosage compensation that allows all but one X chromosome in each cell to be turned off.

Lyon (1961) proposed the single-active X hypothesis to explain the observation that in the mouse, females heterozygous for X-linked fur color genes are patchy mosaics of two colors. To quote Lyon (1961, p. 372): " . . . the evidence of mouse genetics indicates: (1) that the heteropycnotic X-chromosome can be either paternal or maternal in

Giemsa

Fluorescence

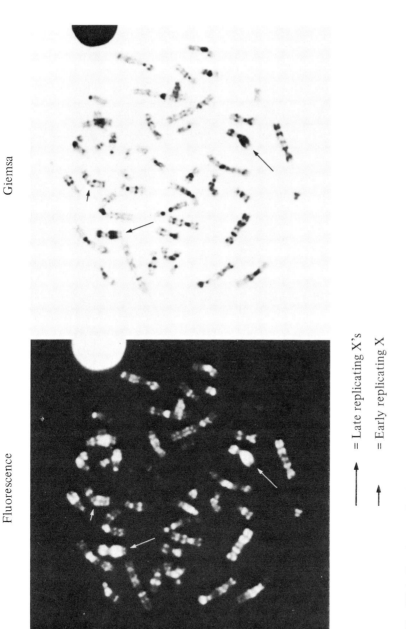

↑ = Late replicating X's

↑ = Early replicating X

Fig. XII.4 Two late-replicating X chromosomes (long arrows) and one early-replicating X chromosome (short arrow) in a woman with 47,XXX (BrdU, left stained with Hoechst 33258, right with Giemsa) (Latt et al, 1976).

origin in different cells of the same animal; (2) that it is genetically inactivated." This hypothesis has inspired an enormous amount of research and the following principles have been established with respect to X-inactivation.

The inactivation of one X chromosome in the normal female, or all but one X in individuals with several X chromosomes, takes place early in embryonic development at an estimated 1000- to 2000-cell stage of the blastocyst or possibly even earlier (cf. Lyon, 1974). The inactivation involves the paternal or maternal X chromosome at random and once an X chromosome is turned off, the event seems to be irreversible. In all the descendants of the cell in which the inactivation took place, the same X is turned off and that one forms the Barr body. The inactive X is not transcribed; it is facultatively heterochromatic.

The single-active X hypothesis implies that a female heterozygous for a gene on the X chromosome should have two different cell populations, each with a different allele being expressed. Studies of patchy fur colors, which have yielded so much information in the mouse, are not possible in human populations. However, in a few human conditions, a patchy appearance resulting from X-inactivation seems to be expressed, for instance, in ocular albinism and in choroidemia (cf. Gartler and Andina, 1976).

On the cellular level, this phenomenon can be studied for a few X-linked genes (cf. Migeon, 1979). The first demonstration of the expression of a single X was done by Davidson et al (1963). They cloned fibroblasts from a woman heterozygous for two allozymes of the enzyme G6PD (glucose-6-phosphate dehydrogenase). On starch-gel electrophoresis, the uncloned cells showed the double pattern, whereas clones of cells revealed only one or the other allozyme.

In a patchy mouse with a normal and an abnormal X chromosome (so-called Cattanach translocation), one X formed the Barr body in one type of patch and the other X formed the Barr body in the other patches. This observation was in accord with the known fur-color genes on the two X chromosomes (cf. Cattanach, 1975). Another building block in the evidence that the late-replicating X chromosome is also the inactive one was provided by observations on mules and hinnies (reviewed by Gartler and Andina, 1976). In these hybrids the donkey-X and the horse-X can be distinguished morphologically. Each species has its own type of G6PD, and the predominant G6PD isozyme in both fibroblasts and erythrocytes of the two hybrids turned out to be the one characteristic of the horse. In agreement with this observation in the two hybrids, the donkey-X was late-replicating in about 90 percent of the cells and the horse-X in 10 percent. Whether this inequality of expression of the two X chromosomes depends on original nonrandomness of the inactivation or originally random inactivation followed by selection for the cell line in which the horse-X is active, is not known.

Behavior of X and Y Chromosomes in Meiocytes

As already pointed out, once X inactivation has taken place it seems to be irreversible. All experimental attempts, whether by treatment with chemicals or hybridization with cells of other mammals, have failed to reactivate the inactive X chromosome.

However, the inactive X is reactivated in the oocytes some time before meiosis, the exact timing being still under dispute (cf. Gartler and Andina, 1976). Both X chromosomes are transcribed and neither shows heteropycnotic behavior. In meiosis the two X chromosomes pair normally and the bivalent formed by them does not differ from autosomal bivalents (Fig. X.4).

Interestingly, in male meiosis the opposite phenomenon occurs. Both the X and Y chromosomes appear heteropycnotic from zygotene to diplotene and are not transcribed during spermatogenesis (cf. Cattanach, 1975).

It seems to be a general characteristic of the sex chromosomes, at least at certain stages of development, that they are out of step with the autosomes; they have a tendency to allocycly. This is correlated with their inactivity during the same stages.

Sex Reversal Genes

It is obvious that, apart from the sex determining genes on the Y chromosome, many other genes, some of them autosomal, affect sexual development in man. In many animal species, such as *Drosophila*, mouse, rat and goat, genes are known that can reverse the chromosomal sex. In man too there are diverse conditions in which an individual develops a phenotype that is the opposite of his or her chromosomal sex. In most cases the cause of such sex reversal is a mutant gene. The different types of XY women have been reviewed by Sarto (1972).

One of the best known of the conditions in which an individual with the XY sex chromosome constitution develops a female phenotype is the testicular feminization syndrome. In the rat and the mouse, a corresponding gene (Tfm) is located on the X. In man too the Tfm gene has recently been found to be X-linked (cf. Migeon et al, 1979).

Families in which the testicular feminization gene is inherited are characterized by the predominance of female births. The affected individuals usually have a fully feminine appearance and often come to the attention of a physician only because of primary amenorrhea (lack of menstruation) and sterility. They have testes in the abdominal cavity or the inguinal canal, but the mutant gene renders the target organs insensitive to androgens produced by the testes.

Other conditions in this group are XY pure gonadal dysgenesis (or

Swyer's syndrome) and the extremely rare gonadal agenesis (Sarto and Opitz, 1973). The representatives of the former group are phenotypic females of greater-than-average height with streak gonads and a resulting lack of secondary sex characteristics. As a consequence, they too suffer from primary amenorrhea and sterility. A phenotypically similar gonadal dysgenesis is also found in women with a normal 46,XX chromosome constitution in whom it is probably caused by another mutant gene. Familial incidence of the last-mentioned syndrome is reported, including one family with four affected daughters (Nazareth et al, 1977).

The XX males have a male phenotype but are sterile (cf. de la Chapelle, 1972). As previously mentioned, the majority of them seems to have a translocated segment from Yp attached to Xp (Evans et al, 1979). The XX males have features in common with XXY males but are shorter than average. The resemblance to Klinefelter's syndrome is not surprising if, in both cases, a gene(s) switches the development of an individual with two X chromosomes into a male groove.

References

Alvesalo L (1978) Tooth sizes in a male with 46,Xdel(Y)(qll) chromosome constitution. IADR Abstr. A. 57

Barr ML, Bertram EG (1949) A morphological distinction between neurones of the male and female, and the behavior of the nucleolar satellite during accelerated nucleoprotein synthesis. Nature 163: 676–677

Breg WR, Genel M, Koo GC, et al (1979) H-Y antigen and human sex chromosomal abnormalities. In: Vallet HL, Porter IH (eds) Genetic mechanisms of sexual development. Academic, New York, pp 272–292

Cattanach BM (1975) Control of chromosome inactivation. Ann Rev Genet 9: 1–18

Chapelle A de la (1972) Analytic review: nature and origin of males with XX sex chromosomes. Am J Hum Genet 24: 71–105

Cohen MM, Shaw MW, MacCluer JW (1966) Racial differences in the length of the human Y chromosome. Cytogenetics 5: 34-52

Cohen MM, MacGillivray MH, Capraro VJ, et al (1973) Human dicentric Y chromosomes: case report and review of the literature. J Med Genet 10: 74–79

Davidson RG, Nitowsky HM, Childs B (1963) Demonstration of two populations of cells in the human female heterozygous for glucose-6-phosphate dehydrogenase variants. Proc Natl Acad Sci USA 50: 481–485

Dronamdraju KR (1965) The function of the Y-chromosome in man, animals, and plants. Adv Genet 13: 227–310

Dutrillaux B (1977) New chromosome techniques. In: Yunis JJ (ed) Molecular structure of human chromosomes. Academic, New York, pp 233–265

Evans HJ, Buckton KE, Spowart G, et al (1979) Heteromorphic X chromosomes in 46,XX males: evidence for the involvement of X-Y interchange. Hum Genet 49: 11–31

Friedrich U, Nielsen J (1974) Pericentric inversion Y in a population of newborn boys. Hereditas 76: 147–152

Gartler SM, Andina RJ (1976) Mammalian X-chromosome inactivation. In: Harris H, Hirschhorn K (eds) Advances in human genetics 7. Plenum, New York

Latt SA (1973) Microfluorometric detection of deoxyribonucleic acid replication in human metaphase chromosomes. Proc Natl Acad Sci USA 70: 3395–3399

Latt SA, Willard HF, Gerald PS (1976) BrdU-33258 Hoechst analysis of DNA replication in human lymphocytes with supernumerary or structurally abnormal X chromosomes. Chromosoma 57: 135–153

Lyon MF (1961) Gene action in the X-chromosome of the mouse *(Mus musculus L.)* Nature 190: 372–373

Lyon MF (1974) Mechanisms and evolutionary origins of variable X-chromosome activity in mammals. Proc R Soc Lond B 187: 243–268

Magnelli NC, Vianna-Morgante AM, Frota-Pessoa O, et al (1974) Turner's syndrome and 46,X,i(Yq) karyotype. J Med Genet 11: 403–406

Migeon BR (1979) X-chromosome inactivation as a determinant of female phenotype. In: Vallet HL, Porter IH (eds) Genetic mechanisms of sexual development. Academic, New York, pp 293–303

Migeon CJ, Amrhein JA, Keenan BS, et al (1979) The syndrome of androgen insensitivity in man: its relation to our understanding of male sex differentiation. In: Vallet HL, Porter IH (eds) Genetic mechanisms of sexual development. Academic, New York, pp 93–128

Mittwoch U (1974) Sex chromatin bodies. In: Yunis JJ (ed) Human chromosome methodology, 2nd edn. Academic, New York, pp 73–93

Nazareth HR de S, Farah LMS, Cunha AJB, et al (1977) Pure gonadal dysgenesis (type XX). Hum Genet 37: 117–120

Ohno S (1979) Major sex-determining genes. Springer, Heidelberg

Ohno S, Nagai Y, Ciccarese S (1978) Testicular cells lysostripped of H-Y antigen organize ovarian follicle-like aggregates. Cytogenet Cell Genet 20: 351–364

Robinson JA, Buckton KE (1971) Quinacrine fluorescence of variant and abnormal human Y chromosomes. Chromosoma 35: 342–352

Rosenfeld RG, Luzzatti L, Hintz RL, et al (1979) Sexual and somatic determinants of the human Y chromosome: studies in a 46,XYp- phenotypic female. Am J Hum Genet 31: 458-468

Sarto GE (1972) Genetic disorders affecting genital development in 46,XY individuals (male pseudohermaphroditism). Clin Obstet Gynecol 15: 183–202

Sarto GE, Opitz JM (1973) The XY gonadal agenesis syndrome. J Med Genet 10: 288–293

Sarto GE, Therman E, Patau K (1974) Increased Q fluorescence of an inactive Xq- chromosome in man. Clin Genet 6: 289–293

Schmid W, D'Apuzzo V (1978) Centromere inactivation in a case of Turner variant with two dicentric iso-long arm Y chromosomes. Hum Genet 41: 217–223

Takagi N, Oshimura M (1973) Fluorescence and Giemsa banding studies in the allocyclic X chromosome in embryonic and adult mouse cells. Exp Cell Res 78: 127–135

Taylor MC, Gardner HA, Ezrin C (1978) Isochromosome for the long arm of the Y in an infertile male. Hum Genet 40: 227–230

Wachtel SS (1979) H-Y antigen and sexual development. In: Vallet HL, Porter IH (eds) Genetic mechanisms of sexual development. Academic, New York pp 271–277

Wolf U (1978) Zum Mechanismus der Gonadendifferenzierung. Bull Schweiz Akad Med Wiss 34: 357-368

Zenzes MT, Wolf U, Engel W (1979) Organization in vitro of ovarian cells into testicular structures. Hum Genet 44: 333–338

XIII
Sex Chromosome Abnormalities

Aneuploidy of X Chromosomes in Individuals with a Female Phenotype

The sex chromosomes show a much wider range of viable aneuploidy than do the autosomes, presumably for the following reasons. On the one hand, the Y chromosome seems to contain very few genes apart from those determining the male sex; on the other, all but one X chromosome in a cell are inactivated, forming X chromatin bodies in the interphase. This rule can be stated another way: there is one active X chromosome for each diploid complement of autosomes.

Fig. XIII.1 summarizes the nonmosaic numerical sex chromosome abnormalities found so far, their incidence in newborn infants of the same sex, and the number of Barr bodies formed by them. One male with XYYYY has also been described, but he had an additional XO cell line (van den Berghe et al, 1968).

Aneuploidy results from nondisjunction in either the meiotic divisions of the parents or the early cleavage divisions of the affected individuals. Aneuploidy for more than one chromosome is the result of nondisjunction in both meiotic divisions, nondisjunction in the meiosis of both parents (which must be a rare coincidence), or abnormalities in more than one mitosis.

The incidence of 47,XXX and 47,XXY children increases with the maternal age, as does that of autosomal trisomies. In these cases nondisjunction apparently occurs mainly in maternal meiosis. The incidence of 45,X children, on the other hand, seems to be independent of maternal age. Indeed, studies on the Xg blood group gene, which is

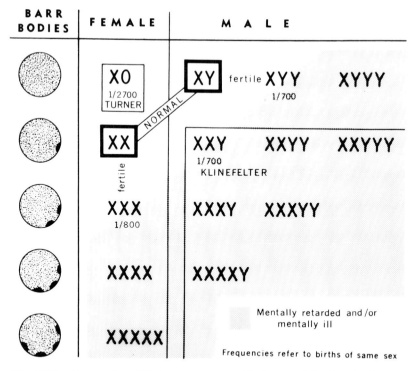

Fig. XIII.1. Human X and Y chromosome constitutions and the Barr bodies formed by the X chromosomes.

located on the X chromosome, show that in 78 percent of the cases, the gamete without a sex chromosome came from the father (Sanger et al, 1971). The syndromes caused by the various sex chromosome aneuploidies have been reviewed by Hamerton (1971) and de Grouchy (1974).

Individuals with a 45,X chromosome constitution develop the so-called Turner's syndrome. They are phenotypic females, but their gonads, which appear normal in the embryo, degenerate postnatally and later consist mainly of fibrous tissue that lacks follicles. Such ovaries do not produce estrogens, and the patients remain sexually infantile. A large number of phenotypic abnormalities has been described in these patients, the most common one being short stature (under 5 ft or 150 cm) (Fig. XIII.2). Other anomalies (Table XIII.1), such as webbed neck, low hairline, shield chest, cubitus valgus (changed carrying angle of the elbow), and pigmented nevi are somewhat less frequent. Often the cardiovascular and urinary systems are also affected.

Patients with Turner's syndrome do not seem to be severely mentally retarded any more often than females with two X chromosomes. Intelligence tests show that their verbal IQs are equal to normal controls, but

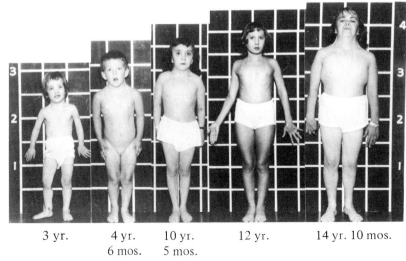

| 3 yr. | 4 yr. | 10 yr. | 12 yr. | 14 yr. 10 mos. |
| | 6 mos. | 5 mos. | | |

Fig. XIII.2. Five patients with Turner's syndrome, showing short stature, characteristic face, wide nose, low set ears, webbed neck, and shield chest (Smith, 1976).

that their performance IQs are lower. This has been interpreted to mean that they have a defect in the cortical part of the right brain hemisphere (cf. Polani, 1977).

An interesting although so far unexplained phenomenon is that about 97 percent of zygotes with a 45,X chromosome constitution end up as spontaneous abortions (Carr, 1971).

Individuals with three X chromosomes do not seem to form a well-

Table XIII.1. Frequency of Turner Symptoms in 332 Individuals with 45,X in Percentage of Those Patients for Whom Symptom Was Recorded. All Ascertainment Types Pooled.[a]

Symptom	%	Symptom	%
Short stature	100	Webbed neck	42
Low hairline	72	Short neck	77
Shield chest	74	Urinary system anomaly	44
Pigmented nevi	64	Hypertension	37
Cubitus valgus	77	Gonadal dysgenesis	91
Short 4th metacarpal	55	Mental retardation	19
Nail anomaly	57	Retarded bone age	64
Cardiovascular anomaly	23	Thyroid disease	18
		Color blindness	11

[a]Denniston and Ulber, unpublished data.

defined syndrome. They are often mentally retarded or psychotic. Their mental status, as well as that of persons with other types of sex chromosome anomalies, is reviewed by Polani (1977). Triple-X women, although fertile, produce overwhelmingly normal children instead of one-half the daughters with 47,XXX and one-half the sons with 47,XXY chromosome constitutions, as might be expected. One possible explanation is that in the first meiotic division of the oocyte, the extra chromosome always ends up in the polar body. Other systems in which aberrant chromosomes are relegated to the polar bodies are found in *Drosophila* and moths.

Patients with more than three X chromosomes suffer from severe mental retardation and have several somatic anomalies. However, their sex development is usually normal. Nielsen et al (1977) reviewed the 26 cases of 48,XXXX found up to that time. They showed a wide range of abnormalities but varied to a considerable extent from case to case. The total of only 26 patients found with this type of aneuploidy shows how rare these multiple sex chromosome anomalies actually are.

As Fig. XIII.1 demonstrates, more than one X chromosome can be found in individuals with a male phenotype. They are discussed in the following paragraphs.

Sex chromosome aneuploidies—mainly XO, XXY, and XXX—are also described in a number of other mammals, such as the mouse, the cat, the dog, and the pig (cf. Lyon, 1974).

Sex Chromosome Aneuploidy with a Male Phenotype

In the group of sex chromosome aneuploidies with a male phenotype, the XXY and XYY conditions are about equally frequent at birth (Fig. XIII.1). Individuals with 47,XXY chromosome constitution form a fairly well-defined, so-called Klinefelter's syndrome. They are characterized by a eunuchoid habitus, their small testes are devoid of germ cells, and they are sterile. They also show a tendency to breast development (gynecomastia). The XXY sex chromosome constitution seems to lower the IQ to some extent, and as a result the affected individuals are often mildly mentally retarded. They also show psychotic tendencies and may eventually end up in correctional institutions (cf. Polani, 1977).

Patients with one Y chromosome and more than two X chromosomes are mentally retarded and display various other symptoms. To take one example, males with the 49,XXXXY chromosome constitution have an IQ in the range of 20 to 50, extensive skeletal anomalies, severe hypogenitalism, strabismus, wide-set eyes, and other anomalies.

The 47,XYY syndrome has unfortunately been sensationalized by the news media, which have often depicted these individuals as a group of murderers and other violent criminals. This simply is not true. What *is* true is that persons with this chromosome complement have a higher probability of coming into conflict with the law than normal males, but their crimes are usually nonviolent (cf. reviews by Hook 1973; Polani, 1977). Most males with an XYY sex chromosome constitution lead normal lives and are not distinguishable from other people. However, they are usually considerably taller than their relatives, often being 6 ft tall and over ($>$ 180 cm).

The highest incidence of XYY males is found when prisoners who are 6 ft tall or taller are chosen for chromosome studies. Apart from their above-average height, XYY males do not represent a defined syndrome, although some neurological symptoms and specific features of body build have been described in them (Daly, 1969). Their fertility does not seem to be impaired to any great extent. This condition is often accompanied by mental retardation.

Males with the chromosome constitution 48,XXYY are also found in increased numbers in prisons. They show features in common with Klinefelter's syndrome, are tall and more or less mentally retarded.

Higher aneuploidies like XXXYY and XYYY are rare. Patients with such sex chromosome constitutions are retarded and display numerous anomalies.

Mosaicism

Mosaicism for sex chromosome aneuploidy seems to be considerably more common than for autosomal aneuploidy. Mosaics with either a 46,XX or 46,XY cell line accompanied by a 45,X cell line are the most common, but three or even more different cell lines have been described. Sometimes the type of mosaicism unequivocally reveals its origin. Thus mosaicism of the type 45,X/46,XX/47,XXX obviously arose by nondisjunction in a chromosomally normal female. In more complicated mosaics—and they can be very complicated, indeed—the mechanisms that generated them remain a matter for speculation.

A wide range of conditions representing intersexes in whom the external genitalia are ambiguous, or hermaphrodites who have gonads containing both testicular and ovarian tissue, has been described (review in Hamerton, 1971). Such persons are often mosaics with the chromosome constitution 45,X/46,XY or 46,XX/46,XY. Whether the individual is more male or more female depends on which cell line is predominant in the different organs. More complicated types of mosaicism are also found in such individuals.

Low-grade mosaicism does not affect the phenotype. As a matter of fact, to some extent all of us are mosaics for chromosomally aberrant cells. Cell lines with 45,X are found in both normal males and females. When the proportion of 45,X cells increases, symptoms of Turner's syndrome begin to appear and we find in this group all the intermediates from normal female phenotype to full Turner's syndrome. Mitotic non-disjunction, as well as other chromosome aberrations, increases in older persons who show significantly more aberrant cells than do younger ones. The proportion of 45,X cells increases especially in older women (cf. Galloway and Buckton, 1978).

In mosaic individuals the same rules apply as in nonmosaics, namely, in each cell all X chromosomes but one are inactivated. If the mosaicism arises after the X inactivation has taken place, only cells in which nondisjunction involves the inactive X chromosome(s) seem to be viable (Daly et al, 1977).

Abnormal X Chromosomes Consisting of X Material

The wide range of aneuploidy displayed by the human X chromosome is matched by a greater variety of structural abnormalities than is found in any of the autosomes. In cells with one normal and one abnormal X chromosome, consisting of X chromosomal material, the abnormal X is always inactivated. This naturally diminishes the effect of the abnormal X. The size of the Barr bodies reflects the sizes of the abnormal X chromosomes. They range from considerably smaller than normal for deleted chromosomes to very large ones for chromosomes consisting of two long arms of the X or even of two almost intact X chromosomes.

Fig. XIII.3 illustrates, apart from X-autosomal or X-Y translocation chromosomes, the main types of abnormal X chromosomes found in man (Therman et al, 1974). The most common of them is the isochromosome for the long arm, i(Xq). An isochromosome in which by definition the two arms are genetically identical may result through different mechanisms. The product of a reciprocal translocation between two X chromosomes is not a true isochromosome, but it cannot be distinguished from one by cytological means. A true but dicentric isochromosome is formed when the two sister chromatids of Xp break near the centromere and join. However, the most common mode of origin is probably misdivision of the centromere (Fig. XI.2), which may also lead to telocentrics of the long arm, tel(Xq).

Deletions of the short or the long arm of the human X chromosome also occur. A deletion of Xp may involve part or all of the arm. In the latter case an acrocentric or telocentric for the long arm is formed. The chromosomes in which the long arm has undergone a deletion range from those resembling a chromosome 18 to metacentrics. Short deletions may be difficult to detect cytologically and they would also cause very few,

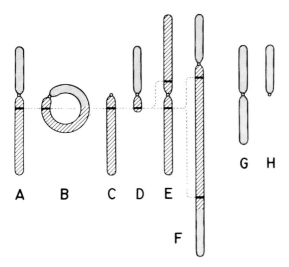

Fig. XIII.3. Five structurally different X chromosomes found in humans (A–E) and two never observed (G–H). Crossline marks presumed inactivation center. A. Normal X; B. ring X; C. telocentric Xq; D. Xq−; E. isochromosome Xq; F. X-X translocation chromosome: attachment long arm-to-long arm; G. isochromosome Xp; H. telocentric Xp (Therman et al, 1974).

if any, symptoms. For the formation of ring chromosomes a distal segment must be missing from both arms to allow the "sticky" ends to join.

Individuals bearing an X chromosome with a deletion are less affected than are those in whom the whole X is missing. Deletions of the Xq cause fewer symptoms, often only amenorrhea, than do deletions of Xp. Few symptoms are displayed by women who have only one-half of the short arm missing. Most of them do not even suffer from gonadal dysgenesis.

In a number of patients ascertained for amenorrhea and showing few other symptoms, a very long chromosome is found in place of one X chromosome. They consist of two X chromosomes attached short arm-to-short arm or long arm-to-long arm with more or less material missing from the two chromosomes. The second centromere in such chromosomes is inactivated, showing no primary constriction. However, a C-band is visible at its site (Fig. VI.1g).

An abnormal X chromosome is often encountered in mosaics in which a second cell line has the chromosome constitution 45,X. In some cases the latter cell line may have come about by the loss of the abnormal X chromosome. In others, both cell types have probably resulted simultaneously from mitotic segregation in a chromatid translocation (Therman and Patau, 1974); this explanation has generally been neglected in the literature.

X Inactivation Center

Whatever the mechanism giving rise to i(Xq) and tel(Xq) chromosomes, short-arm isochromosomes and telocentrics should arise with about the same frequency. However, not one verified case of either has been found. If we compare the relatively frequent abnormal X chromosomes to those that probably do not exist (Fig. XIII.3), it becomes clear that what the former possess and the latter lack is the Q-dark region next to the centromere on the Xq (c region) (Therman et al, 1974; Therman et al, 1980).

In i(Xq) chromosomes and those X-X translocation chromosomes that have this region in duplicate, a certain proportion of Barr bodies is bipartite. From these observations the conclusion was drawn that the c-region (Fig. XII.1a) contains the X inactivation center without which the chromosome cannot form a Barr body, and thus remains active. Two active X chromosomes in the same cell would presumably be a nonviable condition. If two inactivation centers are located on the same abnormal X chromosome, each acts as a center for X chromatin condensation, which leads to the formation of bipartite bodies (Therman et al, 1974).

X-Autosomal Translocations

In more than 30 women, a reciprocal translocation between an X chromosome and an autosome has been found. Those translocation carriers in whom the break in the X has occurred within a certain region of the long arm suffer from primary amenorrhea. The others have a normal phenotype (Sarto et al, 1973). This is possibly the first *position effect,* meaning that the effect of the gene depends on its location, found in man (Therman et al, 1980).

In practically all cases the normal X is inactivated, and the two parts of the other X attached to the autosomal segments remain active. On the other hand, in a female with an unbalanced X-autosomal translocation chromosome, the abnormal chromosome is almost always inactivated; sometimes the inactivation spreads to the autosomal segment, but not always (cf. Therman and Patau, 1974). When the autosomal segment remains active, the phenotype is abnormal, reflecting partial trisomy for the autosome involved.

Such observations have been explained by assuming that at the time of X inactivation, the normal and the abnormal X chromosomes are turned off at random (Gartler and Sparkes, 1963; Sarto et al, 1973). Since, however, inactivation may spread from the X to the attached autosomal segment, cells of a carrier in which translocation chromosomes are inactivated would be badly unbalanced genetically. Intercellular selection would presumably promote the genetically more balanced cell type, which in turn would lead to one inactivation pattern being predominant or exclusive (cf. Therman and Patau, 1974). Using the BrdU technique to study X inactivation, researchers have observed two or more cell lines with different inactivation patterns, one of which, however, is predominant (cf. Therman et al, 1980).

Translocations Involving the Y Chromosome

Most XX men have an extra segment, which presumably originates in Yp, attached to one Xp (Evans et al, 1979). About 10 translocations have been described in which a part of Yq is attached to an X chromosome (van den Berghe et al, 1977; Borgaonkar, 1977). Most of these persons have been phenotypic females with some somatic anomalies, especially short stature. The translocation chromosomes do not show a uniform inactivation pattern, but it varies from random to preferential inactivation of the abnormal one (van den Berghe et al, 1977).

Narahara et al (1978) list 32 Y-autosomal translocation cases. However, those involving the short arms of chromosomes 15 or 22 may not belong to this group, since the satellites of these chromosomes show a bright fluorescence, and this bright material may be duplicated and not come from a translocated Y chromosome.

Abnormal X Chromosome Constitutions and the Phenotype

The abnormal phenotypes of persons with one X chromosome (or part of it) missing or with extra X chromosomes raise a question about why these individuals should differ from normal females and males if one and only one X chromosome is active in every cell. Several hypotheses have been put forward to explain this phenomenon. Therman et al (1980) have reviewed the following hypotheses, which they consider the most plausible.

1) The damage done by abnormal X chromosome constitutions may occur at an early embryonic stage *before* X inactivation.
2) X chromosome abnormalities may exert an effect between reactivation of the inactive X chromosome(s) and meiosis in the oocytes.
3) A recessive gene(s) in a hemizygous condition may be expressed in the cases in which the same X chromosome is active in all cells.

4) A change in the number of presumed active regions on the inactive X chromosome may affect the phenotype.
5) A position effect involving the X chromosome has been established, which means that the region Xq13-q27 has to be intact on both X chromosomes to allow normal female sexual development.
6) A phenotypic effect may be exerted during the selection period by those cells whose inactivation pattern renders them genetically unbalanced.
7) The original X inactivation may be neither regular nor random in all cases with abnormal X chromosome constitutions. Cells with more or less than one active X chromosome would naturally give rise to abnormal phenotypes.

These hypotheses obviously are not mutually exclusive, and the phenotypic effects of a specific X chromosome constitution may simultaneously be exerted through different pathways. Therman et al (1980) came to the conclusion that the idea of a recessive gene has to be discarded and that there is very little evidence for an early embryonic effect. There are some indications but no solid evidence for a nonrandom or irregular X inactivation. The remaining hypotheses (effect of reactivated X chromosomes on the oocytes, change in the number of active chromosome regions on the inactive X chromosomes, position effect, effects during the selection period) probably suffice to explain the phenotypic effects of the various types of X chromosome constitutions.

References

Berghe H van den, Petit P, Fryns JP (1977) Y to X translocation in man. Hum Genet 36:129–141

Berghe H van den, Verresen H, Cassiman JJ (1968) A male with 4 Y-chromosomes. J Clin Endocrinol 28:1370–1372

Borgaonkar DS (1977) Y to X chromosome translocations. Hum Genet 40:113–114

Carr DH (1971) Chromosomes and abortion. In: Harris H, Hirschhorn K (eds) Advances in human genetics, Vol II. Plenum, New York, pp 201–257

Daly RF (1969) Neurological abnormalities in XYY males. Nature 221:472–473

Daly RF, Patau K, Therman E et al (1977) Structure and Barr body formation of an Xp+ chromosome with two inactivation centers. Am J Hum Genet 29:83–93

Evans HJ, Buckton KE, Spowart G et al (1979) Heteromorphic X chromosomes in 46,XX males: evidence for the involvement of X-Y interchange. Hum Genet 49:11–31

Galloway SM, Buckton KE (1978) Aneuploidy and ageing: chromosome studies on a random sample of the population using G-banding. Cytogenet Cell Genet 20:78–95

Gartler SM, Sparkes RS (1963) The Lyon-Beutler hypothesis and isochromosome X patients with the Turner syndrome. Lancet ii:411

Grouchy J de (1974) Sex chromosome disorders. In: Busch H (ed) The cell nucleus, Vol II. Academic, New York, pp 415–436

Grouchy J de, Turleau C (1977) Clinical atlas of human chromosomes. Wiley, New York

Hamerton JL (1971) Human cytogenetics. Clinical cytogenetics, Vol II. Academic, New York

Hook EB (1973) Behavioral implications of the human XYY genotype. Science 179:139–150

Lyon MF (1974) Mechanisms and evolutionary origins of variable X-chromosome activity in mammals. Proc R Soc Lond B 187:243–268

Narahara K, Yabuuchi H, Kimura S, et al (1978) A case of a reciprocal translocation between the Y and no. 1 chromosomes. Jpn J Hum Genet 23:225–231

Nielsen J, Homma A, Christiansen F, et al (1977) Women with tetra-X (48,XXXX). Hereditas 85:151–156

Polani PE (1977) Abnormal sex chromosomes, behavior and mental disorder. In: Tanner JM (ed) Developments in psychiatric research. Hodder and Stoughton, London, pp 89–128

Sanger R, Tippett P, Gavin J (1971) Xg groups and sex abnormalities in people of northern European ancestry. J Med Genet 8:417–426

Sarto GE, Therman E, Patau K (1973) X inactivation in man: a woman with t(Xq−;12q+). Am J Hum Genet 25:262–270

Therman E, Denniston C, Sarto GE, et al (1980) X chromosome constitution and the human female phenotype. Hum Genet 54:133–143

Therman E, Patau K (1974) Abnormal X chromosomes in man: origin, behavior and effects. Humangenetik 25:1–16

Therman E, Sarto GE, Patau K (1974) Center for Barr body condensation on the proximal part of the human Xq: a hypothesis. Chromosoma 44:361–366

XIV
Numerically Abnormal Chromosome Constitutions in Man

Abnormalities of Human Chromosome Number

As described in Chapter II, the chromosome number may change in two different ways. Either the number of *sets* of chromosomes increases, resulting in polyploidy (a decrease, leading to haploidy does not occur in man), or the number of *individual,* normal, chromosomes changes, which gives rise to aneuploidy.

The only types of polyploidy found in man are triploidy and tetraploidy, and the overwhelming majority of these cases end up as spontaneous abortions. The only known autosomal monosomy is the extremely rare 21 monosomy, and of the trisomics only three—those for chromosomes 13, 18, and 21—occur with any appreciable frequency in liveborn children. However, trisomies for most autosomes are found in spontaneous abortions (cf. Boué et al, 1976; Carr and Gedeon, 1977).

Polyploidy

Polyploidy is a common phenomenon in plant evolution and occasional polyploids arise in most plant species. In plants, a tetraploid may be even "bigger and better" than the original diploid, whereas higher polyploids, such as hexaploids and octoploids, are usually less successful.

Polyploidy plays almost no role in animal evolution; individual polyploids occur only occasionally in few animal groups. Among the vertebrates the only major group in which natural polyploidy arises frequently is the amphibians. Especially in salamanders, polyploidy has been found in nature and created experimentally by treating the fertilized eggs with either cold or heat shocks (cf. Fankhauser, 1945). Not only triploid, tetraploid, and pentaploid, but even haploid larvae are viable in newts and axolotls. Although cell size increases with increasing chromosome number, the size and appearance of the larvae from haploid to tetraploid remains about the same. The pentaploids show some abnormalities. Apparently an adjustment of the cell number compensates for increased cell size (Fankhauser, 1945). Among the birds, triploid chickens and turkeys have been described (cf. Niebuhr, 1974).

The situation is very different in mammals. As a rule, polyploidy is lethal and results in prenatal death. Why are the effects of polyploidy so disastrous in this group, when in many other organisms polyploidy is not only harmless but may have beneficial effects? One possible factor is the discrepancy between the chromosome numbers of the maternal and fetal parts of the placenta. This possibility is supported by the observation that in triploid human fetuses the chorion is often abnormal, showing hydatiform changes (cf. Niebuhr, 1974).

Human Triploids

A triploid zygote may come about through various processes (cf. Niebuhr, 1974). Either the egg cell or the sperm may have an unreduced chromosome number, as a result of restitution in either the first or second meiotic division; or the second polar body may reunite with the egg nucleus; or two sperms may penetrate and fertilize the same egg cell.

Triploids form one of the largest groups of *heteroploid* (abnormal chromosome number) spontaneous abortions (Table XIV.1), representing about 17 percent of them (cf. Carr and Gedeon, 1977). An exceedingly small proportion of triploids is born alive and even of these, most die

Table XIV.1. Relative Frequencies of Different Types of Chromosome Anomalies in 1863 Chromosomally Abnormal Spontaneous Abortions.[a]

Chromosome Anomaly	%
Trisomy	52
45,X	18
Triploidy	17
Tetraploidy	6
Other (mainly translocations)	7

[a]Modified from Carr and Gedeon, 1977.

within a day (cf. Niebuhr, 1974). The longest recorded survival time for a triploid infant was five months for an exceptional 69,XXX individual (Cassidy et al, 1977). In such an isolated case, one always suspects hidden mosaicism, and indeed most triploid infants have turned out to be mosaics with a diploid cell line.

Human Tetraploids

Tetraploid zygotes are rarer than triploids among spontaneous abortions. The only apparently nonmosaic tetraploid child reported to be born alive (Golbus et al, 1976) suffered from mental and growth retardation, had multiple anomalies, and lived for about one year. The few other liveborn children have been diploid/tetraploid mosaics. It is not surprising that tetraploid zygotes are rarer than triploids, since tetraploids are more severely affected and there are far fewer mechanisms that give rise to them. The most probable origin is chromosome duplication in a somatic cell at a very early stage of development. Among the other possible origins, it is highly unlikely that a rare unreduced sperm would by chance fertilize an equally rare diploid ovum.

Autosomal Aneuploidy

In a number of plant species, such as corn, barley, spinach, snapdragon (*Antirrhinum*), and the classic object of trisomy studies—Jimson weed (*Datura*), trisomics for all the chromosomes have been created (cf. Burnham, 1962). Each trisomic differs from any of the others and from normal diploid plants in many phenotypic features affecting the mode of growth, the leaves, the flowers, and the fruits. Each chromosome obviously contains genes that, in the trisomic state, modify many organ systems of a plant.

As already mentioned, human trisomics for only three autosomes (13, 18, and 21) are born alive with any appreciable frequency. Trisomy is found for a few others, but rarely. However, trisomics for these three, as well as for most other autosomes, are observed in spontaneous abortions (Table XIV.2) (cf. Boué et al, 1976; Carr and Gedeon, 1977). Mosaics for a normal and a trisomic cell line are found for more autosomes than full trisomy. Apparently a trisomic cell line that would be lethal by itself is sometimes able to exist in a mosaic individual who has a normal cell line too. In addition, an almost infinite variety of abnormal phenotypes is created by partial trisomies for certain chromosome segments distributed among all the autosomes. They are discussed in Chapters XV and XVII.

Table XIV.2. Number of Trisomics for Different Autosomes in a Total of 349 Trisomic Spontaneous Abortions.[a]

Chromosome	1	2	3	4	5	6	7	8	9	10	11
No. of trisomic abortions	0	12	2	4	0	1	14	15	12	10	1
Chromosome	12	13	14	15	16	17	18	19	20	21	22
No. of trisomic abortions	3	10	32	32	104	0	20	1	5	34	37

[a]Data from Boué et al, 1976.

Anomalies Caused by Chromosomal Imbalance

In this book only a few chromosome syndromes are described as examples, since a vast literature dealing with the phenotypes of such conditions already exists. Concise and well-illustrated descriptions of chromosomal and other syndromes are found in Smith (1976). The book by de Grouchy and Turleau (1977) is completely oriented toward clinical cytogenetics. A list of papers dealing with chromosome aberrations has been published by Borgaonkar (1977). Syndrome descriptions are also given in Hamerton (1971), de Grouchy (1974), and Makino (1975). *New Chromosomal Syndromes* (1977) contains reviews on the phenotypes caused by imbalance for whole chromosomes and chromosome segments.

It has been repeatedly stressed that no phenotypic anomaly is exclusive to any chromosome syndrome, but rather that each of them is characterized by a combination of symptoms (cf. Smith, 1977). In the main, this is true, although there are anomalies that are so rare that they can be regarded as characteristic of one or at most a couple of syndromes (cf. Lewandowski and Yunis, 1977). For example, the cat cry in infants with a 5p− deletion is a specific trait. The cat cry syndrome is also characterized by prematurely gray hair. Polydactyly of hands or feet is found in patients with trisomy for chromosome 13, as is persistence of fetal hemoglobin (HbF) and abnormal projections in the neutrophils. Defects of the forebrain (holoprosencephaly) are also typical of 13-trisomic infants. The position of fingers, with the second overlapping the third, and the fifth the fourth, is a characteristic feature of the 18 trisomy syndrome, but is also found in some other syndromes.

In contrast to such specific symptoms, many anomalies are common to most chromosomal syndromes. Thus imbalance for even a small autosomal segment causes mental retardation. This is not surprising since the central nervous system is the most complicated of all organ systems, and therefore even a slight genetic imbalance has a deleterious effect on it.

Growth retardation seems to be almost ubiquitous in chromosomal syndromes, the only exceptions being trisomy for chromosome 8 and for the short arm of chromosome 20 (cf. Francke, 1977). In many chromosomal syndromes, the failure to thrive, which is also reflected in abnormally low birth weight, is so extreme that they are essentially lethal conditions; the few infants who survive birth can be regarded as belated miscarriages (Smith, 1977). For instance, often 18-trisomic infants do not gain any weight after birth.

Heteroploidy seems to affect organ systems according to their complexity. Thus anomalies of the cardiovascular system are characteristic of a great variety of chromosomal syndromes. Ventricular septal defect is found in 47 percent of infants with 13 trisomy, in 34 percent of 18-trisomic patients, and in 20 percent of 21-trisomic children (Lewandowski and Yunis, 1977). A whole spectrum of other heart defects is also present in chromosomal syndromes.

One prerequisite for normal sex development is an XX or XY sex chromosome constitution. However, many autosomal syndromes exhibit defects in this system. For instance, infants with 13 trisomy often show cryptorchidism or a bicornuate uterus. The incidence of such effects is probably underestimated, since many infants with chromosomal syndromes die early, and their mature phenotypes remain unknown.

That no two persons have identical dermatoglyphics indicates that a great number of genes must be involved in their determination. Different constellations of dermatoglyphic features are characteristic of the various chromosomal syndromes. Therefore the genes responsible for dermatoglyphics are probably distributed on many chromosomes. As with other phenotypic anomalies, no single feature is specific for one chromosomal syndrome, but rather each condition is characterized by a combination of features. Individual features associated with specific syndromes also occur occasionally in the dermatoglyphics of normal persons.

The imbalance for even the smallest chromosome segment recognizable under the light microscope, excluding the heterochromatic regions, causes a variety of symptoms affecting many organ systems. This reflects the fact that even a small segment contains hundreds of genes. However, imbalances for *different* chromosome segments of about the *same length* may affect the phenotype to very different degrees. Obviously monosomy or trisomy is much more damaging for some genes than it is for others. It is also possible that the density of genes is higher in the Q-dark segments than in the bright ones, since trisomy or monosomy for them has much more drastic effects (Korenberg et al, 1978).

The phenotypes of the more common chromosome syndromes are well defined by now. However, the mechanisms through which chromosomal imbalance exerts its influence are almost totally unknown. One possible pathway is that cells with different chromosome constitutions do not divide at the same rate (Paton et al, 1974). It is also known that in many mosaics the normal cell line has a selective advantage (cf. Nielsen, 1976). An abnormal division rate of cells may, in turn, change the growth rates of different tissues and organs, which might result in some of the phenotypic anomalies characteristic of chromosomal syndromes. The developmental study of chromosomally abnormal embryos is an interesting field of research that is only beginning (cf. Gropp et al, 1976).

13 Trisomy (D₁ Trisomy, Patau's Syndrome)

The chromosomes of an infant with 13 trisomy were first studied in our laboratory (Patau et al, 1960), and therefore this syndrome is described in detail as an example of trisomy syndromes. The extra D chromosome, which is the cause of the syndrome, was called by Patau et al (1960) D_1 because it was thought that trisomics for the other two D chromosomes might be discovered later. As we now know, this expectation has not been fulfilled, and with banding techniques D_1 has been identified as chromosome 13 (Fig. XIV.1). Most 13-trisomic zygotes end up as spontaneous abortions (Table XIV.2), demonstrating that this condition is semilethal.

The estimates of the frequency of 13 trisomy among newborn infants vary between 1/6,000 and 1/10,000. Increased maternal age is a factor in 13 trisomy, as it is for the other full trisomies. Even those 13-trisomics who survive birth have a limited life expectancy; about 45 percent die within the first month, 90 percent are dead before six months, and fewer than 5 percent reach age 3 years (cf. Gorlin, 1977; Niebuhr, 1977).

Infants with 13 trisomy are severely mentally retarded and often deaf. Various degrees of forebrain defect (holoprosencephaly) are common (Table XIV.3). Eye anomalies range from anophthalmia to microphthalmia (Fig. XIV.2), often combined with coloboma of the iris (fissure of the iris). Cleft lip or cleft palate, or both are also characteristic of this syndrome (Fig. XIV.2). Capillary hemangiomata and scalp defects have often been described. Polydactyly (Fig. XIV.2) was already mentioned. The heel is often prominent and the feet exhibit a rocker bottom. The thumb may be retroflexed (Fig. XIV.2). Different types of heart anomalies are common.

Table XIV.3 gives the frequencies of the most common clinical anomalies in 13 trisomy (Niebuhr, 1977; see also de Grouchy, 1974; Hodes et al, 1978). The incidence of the various symptoms shows a wide range of variation from severe mental retardation, which is always present, to low-set ears, which have been reported in 11 percent of the cases. The considerable phenotypic variation displayed by 13-trisomic infants probably reflects the differences in the allelic content of the three homologous chromosomes. The variation in the symptoms is also caused

Table XIV.3. Frequency of Main Symptoms in 13 Trisomy.[a]

Symptom	% of cases	Symptom	% of cases
Severe retardation	100	Hemangiomata	73
Deafness	53	Polydactyly	78
Microcephaly	59	Heart disease	76
Hypertelorism	93	Renal anomaly	52
Epicanthic folds	52	Bicornuate uterus	43
Microphthalmia or anophthalmia	78	Arhinencephalia (absent olfactory bulbs)	71
Coloboma	35	Simian crease	64
Harelip	55	Distal triradius	77
Cleft palate	65	Elevated fetal hemoglobin (HbF)	>50
Malformed ears	81	Abnormal projections of neutrophils	>50
Low-set ears	11		

[a]Data from Niebuhr, 1977; other authors give somewhat different frequencies.

Fig. XIV.1. Karyotype of a 13-trisomic male (orcein staining). Insert: Q-banded D-group with three chromosomes 13.

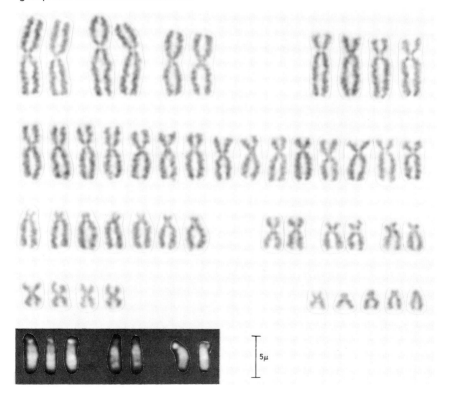

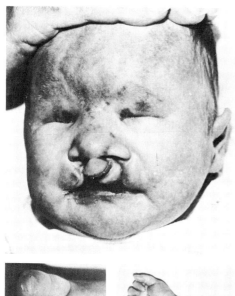

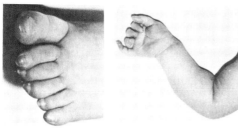

Fig. XIV.2. First D-trisomic child described with anophthalmia, hare lip, six toes, retroflexed thumbs, and hemangiomata (Patau et al, 1960).

by the different ages of the patients, some symptoms vanishing and others becoming apparent with increasing age.

As in other mosaics, a normal cell line dilutes the effect of the trisomic cells, and as a result mosaic individuals have fewer and less severe anomalies than have full trisomics. In most normal trisomy mosaics, the abnormal cell line simply has an extra chromosome 13. However, an effectively trisomy mosaic child with an unusual karyotype was described by Therman et al (1963). In about 45 percent of the cells that had a 46,XX chromosome constitution, one D₁ chromosome was telocentric, whereas in the remaining cells this chromosome was replaced by an isochromosome. It was subsequently shown by autoradiography and by banding that this interpretation was correct. The 4-year-old girl was less retarded and had fewer anomalies than fully trisomic individuals. The isochromosome may have arisen through misdivision from the telocentric or the telocentric from the isochromosome. A similar type of telocentric-isochromosome mosaicism has been observed for a few other chromosomes, especially 21.

Partial trisomy for different parts of chromosome 13 is also found (cf. de Grouchy, 1974; Niebuhr, 1977). As a rule these individuals have only some of the anomalies typical of 13 trisomy; partial trisomy for different segments shows specific combinations of symptoms. The observations of these specific symptoms are used for trisomy mapping of chromosome 13. According to de Grouchy (1974), 80 percent of 13 trisomy patients have the chromosome constitution 47,+13, whereas the rest are either mosaics or have a Robertsonian translocation t(13qDq). The same chromosomes for which *trisomy* is viable are usually viable as *partial monosomics*. This is also true of chromosome 13 in which deletions are used for monosomy mapping (cf. Noel et al, 1976).

21 Trisomy Syndrome (Down's Syndrome, Mongolism)

This condition, which is the least severe of the autosomal trisomy syndromes, is described in detail in the books and reviews previously mentioned; in addition, entire volumes have been dedicated to various aspects of it (cf. Penrose and Smith, 1966; Lilienfeld, 1969). The chromosomal cause of Down's syndrome was discovered by Lejeune et al (1959). The syndrome is by far the most frequent of the autosomal trisomy syndromes, the estimates of its incidence range from 1/500 to 1/1,000 newborns. In roughly 95 percent of the cases the chromosome constitution is 47,+21. In 2 percent of patients ascertained as probable 21 trisomics, mosaicism for a normal and a trisomic cell line is present. However, low-grade 21 trisomy mosaicism, which does not affect the phenotype, is sometimes found, especially in parents of 21-trisomic children. In about 3 percent of the patients the extra chromosome is attached to another chromosome, usually as a result of centric fusion to another acrocentric. Such Robertsonian translocations and their inheritance are discussed in Chapter XVI.

In individuals with 21 trisomy, the probability of developing leukemia is increased 20-fold. Other causes of mortality are heart disease and infections, especially in the respiratory system.

Interestingly, a condition corresponding to 21 trisomy in both symptoms and chromosomal cause is found in the chimpanzee (McClure et al, 1969; Benirschke et al, 1974) and the orangutan (Andrle et al, 1979).

18 Trisomy Syndrome (Edward's Syndrome)

Because of their severe failure to thrive, 18-trisomic infants often appear even more miserable than do 13 trisomics. The many anomalies characteristic of this syndrome have been described by de Grouchy (1974), Gorlin (1977), and Hodes et al (1978). The reported incidence of 18 trisomy varies between 1/3,500 and 1/7,000. Of those born alive, 30 percent die within one month and only 10 percent survive one year (cf. Gorlin, 1977). About 80 percent of the patients have straight trisomy, another 10 percent are mosaics, whereas the rest either are double trisomics for another chromosome or have a translocation (cf. de Grouchy, 1977).

Other Autosomal Aneuploidy Syndromes

Of the other trisomy syndromes, only 22 trisomy has definitely been found in a nonmosaic condition. In their review of this syndrome, Hsu and Hirschhorn (1977) collected 18 cytologically confirmed cases from the literature. Especially interesting is the family studied by Uchida et al (1968), in which the mother was a mosaic and two children had full 22 trisomy. Trisomy for approximately one-half of chromosome 22 also leads to a defined, so-called cat's eye syndrome (cf. Toomey et al, 1977).

Trisomy for chromosome 8 is found repeatedly, but almost always in a mosaic state (Fig. XIV.3). Of the 61 patients reviewed by Riccardi (1977), only one was apparently nonmosaic. Trisomy mosaicism for chromosome 9 has been reported several times, and one case of 10 trisomy has even been described (cf. Makino, 1975). Chromosome 7 is also involved in trisomy mosaicism, and an interesting family in which mother and daughter, both mentally ill, were such mosaics was found by de Bault and Halmi (1975). The most probable explanation for this family is that the zygote from which the daughter developed was trisomic, but later a normal cell line arose through loss of one chromosome 7.

Both mosaic and full 14 trisomy are virtually lethal conditions and only a couple of such infants have been born alive (cf. de Grouchy and Turleau, 1977). Reports of 16 trisomy mosaicism still need to be confirmed. Recently the first case of mosaicism for 3 trisomy was discovered in our laboratory (unpublished). The patient is a 30-year-old severely retarded woman with multiple anomalies. Possibly she had a higher percentage of trisomic cells when she was younger, but during development the normal cells have been selected for. Normal cells now constitute 95 percent of the lymphocytes.

Monosomy in liveborn infants has been established for only one autosome, chromosome 21. This *monosomy 21 syndrome* is extremely rare, and even in spontaneous abortions it is infrequent.

Spontaneous Abortions

Table XIV.1 gives an estimate of the incidence of spontaneous abortions caused by different kinds of chromosome abnormalities. Carr and Gedeon (1977) estimate that 38 percent of spontaneous abortions are heteroploid.

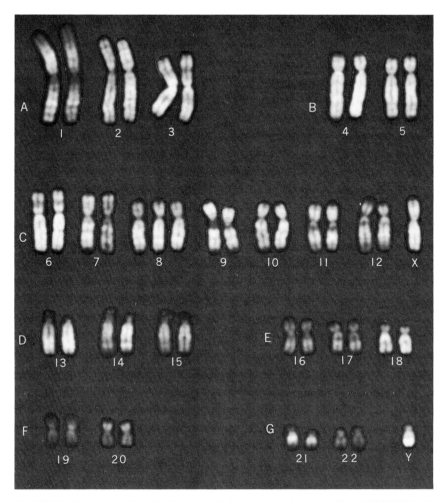

Fig. XIV.3. Karyotype of an 8-trisomic cell from a mosaic male with 46,XY/47,XY,+8 (Q-banding.)

Obviously many factors affect such estimates, which indeed vary greatly in different studies. A number of abortions must occur so early that they are unrecognizable and even of the recognized ones, not all are amenable to chromosome analysis. Early abortions show more chromosome abnormalities than do later ones. The incidence of trisomic abortions increases with the maternal age. The frequency of polyploid as well as of 45,X abortions, on the other hand, is independent of the mother's age (cf. Carr and Gedeon, 1977).

Apart from the pooled trisomic abortions that have been found for most autosomes, polyploidy (mainly triploidy) is the leading chromosomal cause of fetal loss. Of the individual classes, the lack of one sex chromosome (in other words, the chromosome constitution 45,X) is the most frequent. It is estimated that only 3 percent of the XO zygotes are born alive. Obviously the 45,X zygotes fall into separate classes of which one is lethal, whereas those who are born alive are not especially miserable.

Since monosomy is brought about by both chromosome loss and nondisjunction (whereas trisomy results only from the latter process), monosomics should be more frequent among spontaneous abortions than trisomics. In reality, monosomy is almost as rare in recognizable abortions as it is in liveborn children. Obviously monosomy for a chromosome—or for a chromosome segment—has much more deleterious effects than the corresponding trisomy. Apparently the monosomic zygotes, as a rule, die so early that they are not recognized as abortions.

This is also true for trisomic zygotes for certain chromosomes. Thus Boué et al (1976) reported that among 349 trisomic abortions, not one was observed for chromosomes 1, 5, and 17 (Table XIV.2) and even among the other chromosomes, trisomic abortions were distributed very unevenly. As mentioned in Chapter VI, Korenberg et al (1978) have put forward the hypothesis that chromosomes with especially gene-rich segments act as trisomy lethals (Patau, 1964) at a very early stage of gestation.

Another factor that may bias the observed numbers is the time at which the trisomic condition is lethal. For instance, the time of death caused by trisomy 13 ranges from early embryonic development to several years postnatally, whereas the lethal effect of chromosome 16 apparently occurs within a very limited period of gestation; the reported times vary from 22 to 31 days, at which stage chromosome studies are feasible (cf. Boué et al, 1976).

References

Andrle M, Feidler W, Rett A, et al (1979) A case of trisomy 22 in *Pongo pygmaeus*. Cytogenet Cell Genet 24:1–6

Bault LE de, Halmi KA (1975) Familial trisomy 7 mosaicism. J Med Genet 12:200–203 (1975)

Benirschke K, Bogart MH, McClure HM, et al (1974) Fluorescence of the trisomic chimpanzee chromosomes. J Med Prim 3:311–314

Borgaonkar DS (1977) Chromosomal variation in man: a catalog of chromosomal variants and anomalies, 2nd ed. Liss, New York

Boué J, Daketsé M-J, Deluchat C, et al (1976) Identification par les bandes Q et G des anomalies chromosomiques dans les avortements spontanés. Ann Genet 19:233–239

Burnham CR (1962) Discussions of cytogenetics. Burgess, Minneapolis

Carr DH, Gedeon M (1977) Population cytogenetics of human abortuses. In: Hook EB, Porter IH (eds) Population cytogenetics. Academic, New York, pp 1–9

Cassidy SB, Whitworth T, Sanders D, et al (1977) Five month extrauterine survival in a female triploid (69,XXX) child. Ann Genet 20:277–279

Fankhauser G (1945) The effects of changes in chromosome number on amphibian development. Q Rev Biol 20:20–78

Francke U (1977) Abnormalities of chromosomes 11 and 20. In: Yunis JJ (ed) New chromosomal syndromes. Academic, New York, pp 245–272

Golbus MS, Bachman R, Wiltse S, et al (1976) Tetraploidy in a liveborn infant. J Med Genet 13:329–332

Gorlin RJ (1977) Classical chromosome disorders. In: Yunis JJ (ed) New chromosomal syndromes. Academic, New York, pp 59–117

Gropp A, Putz B, Zimmermann U (1976) Autosomal monosomy and trisomy causing developmental failure. In: Gropp A, Benirschke K (eds) Current topics in pathology 62. Springer, Heidelberg, pp 177–192

Grouchy J de (1974) Clinical cytogenetics: autosomal disorders. In: Busch H (ed) The cell nucleus. Academic, New York, pp 371–414

Grouchy J de, Turleau C (1977) Clinical atlas of human chromosomes. Wiley, New York

Hamerton JL (1971) Human cytogenetics, Vol II. Academic, New York

Hodes ME, Cole J, Palmer CG, et al (1978) Clinical experience with trisomies 18 and 13. J Med Genet 15:48–60

Hsu LYF, Hirschhorn K (1977) The trisomy 22 syndrome and the cat eye syndrome. In: Yunis JJ (ed) New chromosomal syndromes. Academic, New York, pp 339–368

Korenberg JR, Therman E, Denniston C (1978) Hot spots and functional organization of human chromosomes. Hum Genet 43:13–22

Lejeune J, Turpin R, Gautier M (1959) Le mongolisme, premier example d'aberration autosomique humaine. Ann Genet 1:41–49

Lewandowski RC, Yunis JJ (1977) Phenotypic mapping in man. In: Yunis JJ (ed) New chromosomal syndromes. Academic, New York, pp 369–394

Lilienfeld AM (1969) Epidemiology of mongolism. Johns Hopkins, Baltimore

Makino S (1975) Human chromosomes. Igaku Shoin, Tokyo

McClure HM, Belden KH, Pieper WA, et al (1969) Autosomal trisomy in chimpanzee: resemblance to Down's syndrome. Science 165:1010–1012

Niebuhr E (1974) Triploidy in man: cytogenetical and clinical aspects. Humangenetik 21:103–125

Niebuhr E (1977) Partial trisomies and deletions of chromosome 13. In: Yunis JJ (ed) New chromosomal syndromes. Academic, New York, pp 273–299

Nielsen J (1976) Cell selection in vivo in normal/aneuploid chromosome abnormalities. Hum Genet 32:203–206

Noel B, Quack B, Rethore MO (1976) Partial deletions and trisomies of chromosome 13; mapping of bands associated with particular malformations. Clin Genet 9:593–602

Patau K (1964) Partial trisomy. In: Fishbein M (ed) Second international conference on congenital malformations. International Medical Congress, New York, pp 52–59

Patau K, Smith DW, Therman E, et al (1960) Multiple congenital anomaly caused by an extra autosome. Lancet i:790–793

Paton GR, Silver MF, Allison AC (1974) Comparison of cell cycle time in normal and trisomic cells. Humangenetik 23:173–182

Penrose LS, Smith GF (1966) Down's anomaly. Little, Brown, Boston

Riccardi VM (1977) Trisomy 8: an international study of 70 patients. In: Birth defects: original article series, XIII, No. 3C. The National Foundation, New York, pp 171–184

Smith DW (1976) Recognizable patterns of human malformation, 2nd edn. Saunders, Philadelphia

Smith DW (1977) Clinical diagnosis and nature of chromosomal abnormalities. In: Yunis JJ (ed) New chromosomal syndromes. Academic, New York, pp 55–58

Therman E, Patau K, DeMars RI, et al (1963) Iso/telo-D_1 mosaicism in a child with an incomplete D_1 trisomy syndrome. Portugal Acta Biol 7:211–224

Therman E, Patau K, Smith DW, et al (1961) The D trisomy syndrome and XO gonadal dysgenesis in two sisters. Am J Hum Genet 13:193–204

Toomey KE, Mohandas T, Leisti, et al (1977) Further delineation of the supernumerary chromosome in the cat-eye syndrome. Clin Genet 12:275–284

Uchida IA, Ray M, McRae KN, et al (1968) Familial occurrence of trisomy 22. Am J Hum Genet 20:107–118

Yunis JJ (ed) (1977) New chromosomal syndromes. Academic, New York

XV
Structurally Abnormal Human Autosomes

Structurally Abnormal Chromosomes

In contrast to full trisomy and monosomy, which in liveborn infants are limited to a few autosomes, the variety of structurally abnormal chromosomes is almost infinite. This is to be expected since chromosomes may break at almost any point, and the broken ends may join randomly to form new combinations. Chromosome breaks occur in meiocytes as well as in somatic cells. The only limitation to the variety of structurally abnormal chromosomes is their possible lethal effect on the individuals or cells carrying them.

It would be impossible to review the entire field of structural chromosome abnormalities in man even in a much more comprehensive cytogenetics book than this one. Therefore only the main classes of abnormal chromosomes are described herein, illustrated by a few examples of the resulting syndromes. Reciprocal and Robertsonian translocations and their segregation are discussed in separate chapters.

Chromosomal Polymorphisms

Chromosomal polymorphisms are structural variants of chromosomes that are widespread in human populations and have no effect on the phenotype, even in their most extreme forms. The apparent harmlessness of chromosomal polymorphism led to the conclusion that segments displaying such variants must be heterochromatic. With banding techniques we now know that these polymorphisms represent constitutive heterochromatin, often at the centric regions of the chromosomes. In a comparison of the incidence of such chromosome variants in 200 mentally retarded patients and the same number of normal controls, no difference was found between the two groups (Tharapel and Summitt, 1978).

The following chromosome segments display polymorphism: (1) The Q-bright distal end of the Y chromosome, which varies from zero to three or four times its average length (Fig. XII.1b), extreme variants being very rare. (2) The short arms and satellites of acrocentric chromosomes which vary both in size and in fluorescent characteristics. The entire short arm may be missing, whereas on the other hand double tandem satellites have been observed. The variation of the satellites and short arms is even more difficult to quantify than that of other heterochromatic regions. (3) A Q-bright centric region may or may not be present in chromosomes 3, 4, 13, and 22. (4) Centric heterochromatin as revealed by C-banding shows a wide range of variation, especially in chromosome 1 (Fig. VI.1f), 9, and 16, but some variation is seen in many other chromosomes, such as 12, 17, and 21 (Mayer et al, 1978) and 19 (Trunca Doyle, 1976). It is likely that all the C-bands vary to some extent (cf. Craig-Holmes, 1977) (Fig. XV.1).

Possibly the variation of the distal segment of the Y and of the C-bands in chromosomes 1, 9, and 16 is not continuous, but the heterochromatin consists of "blocks" whose number may vary (Madan and Bobrow, 1974; Magenis et al, 1977).

The most probable mechanism creating the variation in the heterochromatic segments is thought to be unequal exchanges in mitosis, since constitutive heterochromatin consists of repeated sequences of DNA. On the other hand, meiotic crossing-over rarely, if ever, occurs in constitutive heterochromatin (cf. Kurnit, 1979). However, even unequal mitotic exchanges must be relatively rare, since heterochromatic variants show straightforward mendelian inheritance. Several families in which a change in heterochromatin has probably occurred are reported by Craig-Holmes (1977).

In addition to changes in size, heterochromatic regions are prone to other structural aberrations. Numerous inversions that involve the centric heterochromatin of chromosomes 1 and 9 have been described (cf. Hansmann, 1976; Magenis et al, 1977). In a highly inbred family, two members were even found to be homozygous for a chromosome 9 inversion (Vine et al, 1976). Inversions involving the bright centric band in chromosome 3 have also been described (Soudek and Sroka, 1978).

Quantitative Comparisons

Quantitative comparisons of the sizes of C-bands are difficult, even within one study. This is also true of determining inversions in these regions. It is therefore not surprising that the estimates of the incidence

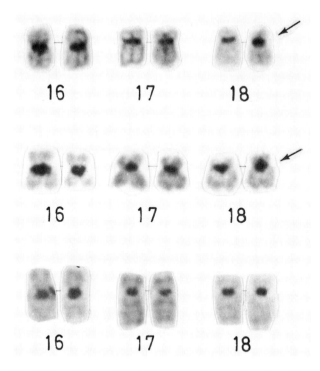

16   17   18

16   17   18

16   17   18

Fig. XV.1. E group from father, daughter, and mother. The exceptionally large C-band in chromosome 18 in father and daughter was used for determination of paternity (courtesy of HD Hager and TM Schroeder).

of heterochromatic variants differ wildly. For instance, reports on the frequency of heterochromatic size variants in chromosome 9 range from 0.1 percent to 12.5 percent (cf. Sanchez and Yunis, 1977), and the frequency of inversions involving the same region from 0.7 percent to 11.3 percent (cf. Sanchez and Yunis, 1977). Although populations may indeed vary with respect to the incidence of heterochromatic variants, clearly the strikingly different results are mainly caused by the nonuniform criteria used, combined with technical difficulties. The final results on the incidence of heterochromatic variants are not yet in.

A few other chromosome variants whose character is still largely unclear have been described. For instance, the middle of 17p sometimes shows a secondary constriction. So-called fragile regions are another form of rare variant; they usually appear as unstained gaps in the chromosome and have a tendency to break. Such regions have been described in several chromosomes (Chapter VII).

Heterochromatic variants are used as linkage markers in gene mapping and as markers in cell hybridization studies. They can also be helpful in determining the heterozygosity of twins and in settling disputes about paternity (Fig. XV.1). Studies have also been done to determine in which parent the aberration occurred that led to the birth of a trisomic or triploid infant.

Inversions

Inversions represent another type of structural chromosome change that does not affect the phenotype of the carrier. Before banding techniques came into use, only pericentric inversions that shifted the position of the centromere could be detected. However, even with banding, pericentric inversions are detected much more frequently than are paracentric ones. Indeed the author is aware of only three published cases of paracentric inversions, two of them in 7q and one in 5q (Shimba et al, 1976; Canki and Dutrillaux, 1979). Strangely enough, as Canki and Dutrillaux point out, their chromosome 7 inversion seems to restore its structure to that found in the chimpanzee.

As already mentioned, centric heterochromatin is often involved in small inversions. However, large inversions are reported in most human chromosomes, the list now including 1–5, 7–10, 13–15, 18, 19, the X, and the Y chromosomes (cf. Borgaonkar, 1977). Inversions involving chromosome 2 have been reported a number of times, and this chromosome may be especially prone to them (cf. Leonard et al, 1975; Phillips, 1978). Interestingly, Leonard et al (1975) found in four unrelated families an inv(2) in which the breakpoints seemed identical. The breakpoint on 2q was in a region that has repeatedly been found to be a fragile site.

In meiosis, short heterozygous inversions usually remain unpaired, whereas a larger inverted segment forms a loop to pair with its homologue. Crossing-over in a loop leads to a deletion and duplication (Fig. XV.2), a phenomenon called "aneusomie de recombinaison" by the French authors. Crossing-over in an inversion may produce a dicentric and an acentric chromatid, and the presence of heterozygous inversions in various organisms has often been inferred from the resulting bridges and fragments in meiosis. They are especially frequent in the many plants that are capable of asexual reproduction. The complicated relationships of various types of crossing-over in inversions and the chromosome configurations arising from them are reviewed in many cytogenetics textbooks, for instance, in Burnham (1962).

It is surprising how few of the inversions in man seem to lead to reproductive trouble—either to the birth of recombinant offspring or to partial sterility (cf. Moorhead, 1976). For small inversions this is understandable since they rarely undergo legitimate pairing. However, even larger inversions are often genetically benign.

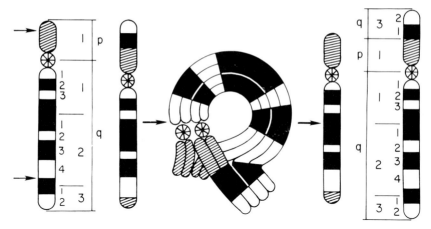

Fig. XV.2. Breakpoints in a pericentric inversion of chromosome 14 and two types of abnormal chromosomes (one with duplication, the other with deletion) resulting from crossing-over in an inversion (Trunca and Opitz, 1977).

Trunca and Opitz (1977) in a study of a woman with inv(14) and her abnormal child who had a duplication in the same chromosome (Fig. XV.2), review the factors promoting the incidence of abnormal offspring. The risk for an inversion carrier is determined by the probability of either type of recombinant offspring surviving birth. Trunca and Opitz (1977) divided the pericentric inversions into those involving less than one-third of the chromosome and those involving more. All families in the latter group had at least one abnormal child, those in the former none. The obvious explanation is that the longer the inversion is relative to the chromosome, the greater the probability is that crossing-over will occur in it. At the same time, the duplicated and deleted segments become smaller, as less of the chromosome remains outside the inverted segment. Both factors promote the birth of abnormal but viable offspring.

Of recombinant offspring, the type with the larger duplication and smaller deletion is usually more viable. However, in some unfortunate families both types of abnormal children have been born alive (for example, Vianna-Morgante et al, 1976).

Significant inversions (those not involving heterochromatic regions) are much rarer events than reciprocal translocations. This is to be expected, since the probability of two breaks taking place in the same chromosome is naturally smaller than if the breaks occurred in two chromosomes. Two breaks in the same chromosome arm is an even rarer phenomenon, which accounts for the scarcity of paracentric inversions.

Deletion or Pure Monosomy

Chromosome *deletions* may be divided into two groups, pure deletions and deletions involving reciprocal translocations. Banding techniques reveal that a surprising variety of significant deletions is compatible with human life. Deletions affecting most chromosome ends as well as many interstitial segments have been found in liveborn infants.

As with sex chromosome anomalies, monosomy for an entire autosome or segment thereof has much more serious phenotypic consequences than the corresponding trisomy. Monosomy syndromes also display a wider range of phenotypic variation than those caused by trisomy. A good example of such variation is provided by the family described by Uchida et al (1965). A retarded, apparently nonmosaic woman with a deletion of the whole 18p had two children with the same chromosome anomaly. The viability of the mother and the first child did not seem to be impaired, although they displayed a number of anomalies. The second child, however, who was much more severely affected died on the third day after birth.

Cri du Chat (Cat Cry) Syndrome

By far the most common of the deletion syndromes, which also has been most extensively studied, is the so-called cri du chat syndrome, caused by a partial deletion of 5p. The incidence of this condition in infants is estimated as 1/45,000 and its frequency among the mentally retarded as 1.5/1000 (cf. Niebuhr, 1978b).

Niebuhr (1978b) recently reviewed 331 patients with cri du chat syndrome. One of the most consistent symptoms is the catlike infant cry which, however, is modified with age. The cry seems to be caused mainly by defects of the central nervous system (Schroeder et al, 1967). Niebuhr (1978b) lists 50 symptoms characteristic of this syndrome, many of which are probably interdependent. In addition to the characteristic cry, the most common abnormalities according to Smith (1976) are: severe mental retardation (100 percent), the IQ in adults being less than 20; hypotonia (100 percent), which in adults turns into hypertonia; microcephaly (100 percent); round face in children (68 percent); hypertelorism (94 percent); epicanthic folds (85 percent); downward slanting of palpebral fissures (81 percent); strabismus (61 percent); low-set or poorly developed ears (58 percent); heart disease of various types (30 percent); and characteristic dermatoglyphics (about 80 percent).

In Fig. XV.3 three cri du chat patients representing different age groups are portrayed. The difference in the severity of corresponding monosomy and trisomy syndromes is well demonstrated by cases in

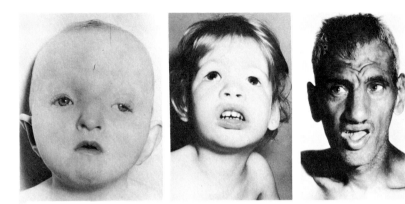

Fig. XV.3. Three cri du chat patients (infant, 4 years, 40 years), showing moon face, hypertelorism, antimongoloid slant of the eyes, downward slant of the mouth, and low-set ears (courtesy of R Laxova).

which segregation in translocation families has produced both cri du chat and its "countertype" offspring, who have partial trisomy for 5p. As a rule, the latter have fewer symptoms, suffering mainly from mental retardation (cf. Yunis et al, 1978).

In a family described by Opitz and Patau (1975), two carriers had a t(5p−;12q+). Six infants with numerous anomalies and partial trisomy for 5p were born in two generations. The severity of the syndrome and the lack of any cri du chat offspring probably depend on most of 5p being deleted, as shown by the diplochromosome karotype of a carrier (Fig. XV.4).

The size of the deletion of 5p causing the cri du chat syndrome varies from very small to about 60 percent of the length of the chromosome arm. It was observed years ago that the length of the deletion showed little correlation with the severity of the syndrome. This became understandable when Niebuhr (1978a) in a detailed cytological study of 35 patients, using both Q-banding and R-banding, showed that the critical segment (whose absence caused most of the symptoms) was a very small region around the middle of 5p15 (Fig. XV.5). Twenty seven of the 35 patients apparently had a terminal deletion, whereas four were the result of translocations, two of which were familial. Similar results have been obtained in larger populations (cf. Niebuhr, 1978b): in 80 percent of the patients the syndrome was caused by a deletion, in 10 percent one of the parents had a translocation, whereas another 10 percent of the patients showed other chromosome abnormalities, such as rings, de novo translocations, and mosaicism. All in all, 12 percent of the cases were familial.

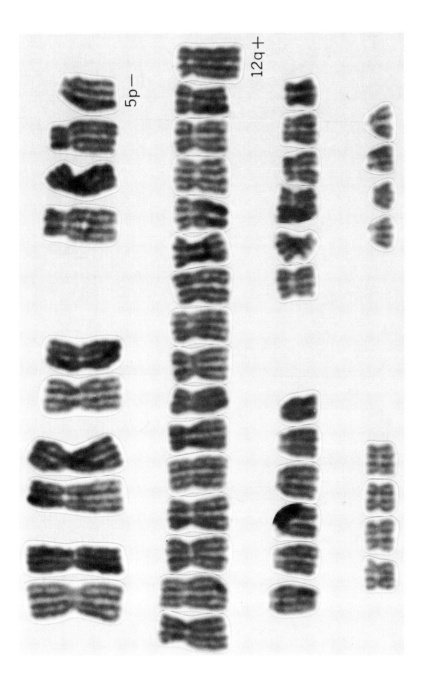

Fig. XV.4. Diplochromosome karyotype of a balanced carrier of t(5p−;12q+) (confirmed with Q-banding).

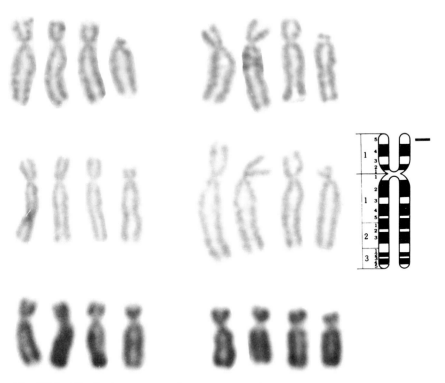

Fig. XV.5. The B chromosomes from two cells each of three cri du chat patients; length of the deletion is different in each patient. The lack of middle part of 5p15 (marked in diagram) is supposed to cause all the symptoms.

Other Deletion Syndromes

In addition to the cri du chat syndrome, diverse other established syndromes are caused by simple deletions. Many other deletions are involved in partial trisomy-monosomy syndromes resulting from translocations. A deletion of 4p is the cause of the so-called Wolf-Hirschhorn syndrome, which has been reviewed on the basis of more than 40 cases by Johnson et al (1976; see also Gorlin, 1977). Defined syndromes also result from deletions of 18p (cf. Schinzel et al, 1974) and of 18q (cf. Trunca Doyle, 1976). Kunze et al (1972) have compared the syndromes caused by 18p−, 18q−, and r(18) chromosomes. Other partial deletions that, although rarer, have also led to the establishment of defined syndromes (cf. de Grouchy and Turleau, 1977) are: 9p− (cf. Alfi et al, 1976; Rethoré, 1977), 11q− (cf. Francke, 1977) 12p− (cf. Rethoré, 1977), and 22q− (cf. Hsu and Hirschhorn, 1977). Isolated cases of partial deletions have been described for most other chromosome arms.

Ring Chromosomes and Their Phenotypic Effects

Since a *ring* involves a deletion at each end of the chromosome, the resulting phenotypes overlap with deletion syndromes for the same chromosome. However, it has been observed repeatedly that the phenotypes of patients with what looks like the same ring chromosome vary greatly. Sometimes two or more patients with apparently identical ring chromosomes have nothing in common phenotypically (for example, Zackai and Breg, 1973; Dallapiccola et al, 1977).

Several factors are responsible for this phenomenon. In earlier textbooks, ring chromosomes were described as either stable or unstable, and indeed some rings show a much greater stability than others. However, since the ubiquitous presence of sister chromatid exchanges has been established, one wonders why any ring chromosome should show even a slight degree of stability. One reason for this phenomenon is the low rate of naturally occurring sister chromatid exchanges and the nonrandomness of their location. Consequently they may be extremely rare in certain chromosomes or chromosome segments. Also the viability of cell lines deviating from the original one may vary greatly for different chromosomes. Perhaps factors like these create the illusion of a stable ring.

Following the occurrence of one sister chromatid exchange, a continuous double ring with one centromere is found in the next metaphase. When the centromere divides in anaphase, the daughter centromeres may go to the same pole, leading to a double, dicentric ring in one daughter cell and no ring in the other; or if the centromeres go to opposite poles, the daughter cells may obtain unequal rings. If the chromosome is twisted when the chromatids join, a new dicentric will be formed. It is easy to see how such mechanisms can give rise to an almost infinite variety of derivatives of the original ring. The main variants are: rings with different numbers of centromeres (even an octocentric ring in a polyploid cell has been described by Niss and Passarge, 1975), interlocking rings, rings consisting of variable combinations of chromosome segments, rings that have opened up, interphase-like chromosomes in metaphase, and several rings in one cell (cf. Hoo et al, 1974).

It is interesting that in many ring chromosome carriers, cells or cell lines in which the ring is missing have been observed. However, what is unexpected is that cell lines which are monosomic for certain chromosomes, for instance 6 (van den Berghe et al, 1974) or 10 (Lansky et al, 1977) are viable, although they have never been observed in persons without a ring chromosome, One cannot help wondering whether the cells with double rings might somehow compensate for the monosomic cells.

Phenotypic Variation

Causes of the great phenotypic variation in carriers of rings involving the same chromosome include: (1) the size of the original deletions, (2) the rate of sister chromatid exchanges in the ring, and (3) the viability of the cell lines with aberrant ring chromosomes.

The deletions involved in the formation of a ring may be very small; sometimes only the telomeres appear to be missing (Fig. XV.6). If in addition the ring is fairly stable, the carrier may be surprisingly normal. Thus a 5-year-old girl with an r(4) was not even mentally retarded. However, she displayed short stature, a small head, and retarded bone age (Surana et al, 1971). More puzzling is the case of a nonretarded 9-year-old girl described by Lansky et al (1977) who had very few other anomalies either. One-half of her lymphocytes showed an r(10), whereas the other half were monosomic for chromosome 10. Since only lymphocytes were studied, she may have been a mosaic with a normal cell line, which is also the most probable explanation for other cases that deviate from the expectation. Undetected mosaicism may also have been present in a 10-year-old girl with r(17) (Ono et al, 1974) since this chromosome is practically never otherwise involved in abnormalities.

Figures XV.6 and VII.2g illustrate an r(9) in which the deletions obviously were minute. However, the severe retardation and multiple congenital anomalies in this patient (originally she had been diagnosed as having the Cornelia de Lange syndrome) are easy to understand (Motl and Opitz, unpublished), as roughly 25 percent of her cells had a double or otherwise abnormal ring chromosome. Her phenotype falls into the — admittedly variable — range of conditions found in carriers of an r(9) (cf. Nakajima et al, 1976; Inouye et al, 1979).

Most human chromosomes are found as rings: 1, 3, 4–7, 9–11, 13, 14 (doubtful), 15–22, and the X and Y chromosomes (cf. Borgaonkar, 1977). In agreement with other chromosome abnormalities, the ring chromosomes are most common for the autosomes for which trisomy is viable, 13, 18, and 21. Rings for chromosomes 5 and 9 are also relatively frequent.

Since carriers of ring chromosomes, apart from the two exceptions already mentioned, are retarded, often severely so, this chromosome abnormality is practically never inherited but arises each time de novo. However, in one case a retarded and psychotic woman with a r(18) gave birth to a daughter bearing the same chromosome who became even more retarded and psychotic. In addition, the child's father suffered from schizophrenia (Christensen et al, 1970).

In all the cases herein discussed, the ring chromosome replaced a member of the normal chromosome complement. However, numerous patients have been described who had a small extra ring chromosome. Such small rings may disappear, leading to mosaicism, or they may open

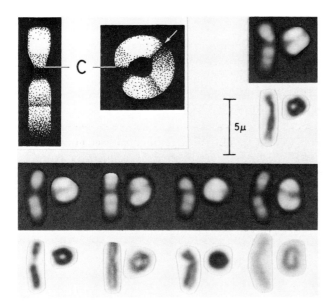

Fig. XV.6. Ring chromosome 9 with very small deletions and the normal 9 and r(9) from 5 cells; C marks centromere (courtesy of ML Motl).

up. The origin of an extra ring chromosome is usually impossible to determine.

Insertions

While deletions require only one or two breaks and inversions two, *insertions* are necessarily the result of three breaks. Consequently they are considerably rarer than abnormalities involving one or two breaks. An insertion may occur either between two chromosomes or within one chromosome, and the segment may be inserted straight or in an inverted position.

As an example, let us take the interesting family described by Therkelsen et al (1973). In the carrier father a segment from 2p had been inserted into 2q. Two abnormal infants were born, representing different types of recombination products.

In another family in which a segment from chromosome 13 had been inserted in an inverted position into 3q, segregation could be followed in three generations (Toomey et al, 1978). Both the monosomic and the trisomic types of recombinant individuals, as well as carriers and normals, were observed.

Duplication or Pure Partial Trisomy

Partial trisomy that is not combined with a deletion either for the same chromosome or for a nonhomologous one is relatively rare. However, such cases are of special interest in human cytogenetics, since they are more convenient for trisomy mapping than are translocation chromosomes in which deletion symptoms interfere with the analysis.

In a couple of cases, an abnormal individual has been found to have a de novo duplication (Vogel et al, 1978). Either of two mechanisms would give rise to such a tandem duplication: an insertion from the homologous chromosome or unequal crossing-over in meiosis or mitosis.

As a rule, crossing-over in an inversion leads to abnormal chromosomes that have both a duplication and a deletion. Only when the inversion involves an acrocentric chromosome and one break has taken place in the short arm will the recombinant chromosome have a pure duplication (Fig. XV.2) (Trunca and Opitz, 1977).

One of the clearest examples of pure partial trisomy is provided by isochromosomes. However, such chromosomes are very rarely members of a 46-chromosome complement, since this would involve monosomy for the other chromosome arm. In consequence, apart from Xq and Yq, isochromosomes have been found only for 9p and 18q (Rodiere et al, 1977), and the acrocentrics 13 and 21, in which the lack of the short arm is not deleterious.

Isochromosomes in addition to a normal chromosome complement probably exist for very short arms, such as 18p, although with the best of banding methods it is impossible to identify them unambiguously (cf. Nielsen et al, 1978). In such cases the individual is tetrasomic for the arm concerned. A chromosomally interesting child was described by Leschot and Lim (1979). She had a t(2q;5q), whereas 5p formed an isochromosome, which resulted in pure trisomy for this arm. A similar combination of translocation and isochromosome formation for 4p has also been found (André et al, 1976).

Small chromosomes that are extra to the normal chromosome complement are found both in normal and abnormal individuals. Mosaicism is common in them, and with increasing age the proportion of the cells with the extra chromosome often decreases. Although in most cases the exact origin of the small extra chromosomes remains uncertain, they can be divided into different categories. (1) Metacentric chromosomes with satellites at both ends, which do not affect the phenotype, obviously consist of two heterochromatic short arms of the acrocentrics. (2) The extra chromosome represents a deleted acrocentric chromosome with satellites on its short arms, or sometimes on both arms. They have been established for all the acrocentrics (cf. Hsu and Hirschhorn, 1977; Niebuhr, 1977; Wyandt et al, 1977). (3) The extra chromosome consists of the centric region of a nonacrocentric chromosome. Obviously two factors determine the size of such chromosomes. On the one hand, very small chromosomes are usually lost; on the other, the longer the chromosome is, the more likely it is to have lethal effects. (4) An extra chromosome may be the result of an 1:3 segregation in a translocation carrier. (5) Isochromosomes have already been discussed. (6) As previously mentioned, small rings occur in addition to the normal chromosome complement. Sometimes they open up and resemble ordinary chromosomes.

Trisomy for 9p is caused by either an isochromosome for this arm or an extra free 9p chromosome (cf. Biederman, 1979). Even tetrasomy for major parts of chromosome 9 seems to be compatible with at least a limited life span. The largest partial tetrasomy has been described in a patient with an extra chromosome consisting of two chromosomes 9 attached long arm-to-long arm (breakpoint in both q22) with the second centromere partly inactivated (Wisniewski et al, 1978). The infant was highly abnormal and lived for only a couple of hours. Four other cases of partial tetrasomy for the same chromosome are reviewed by Wisniewski et al (1978).

References

Alfi OS, Donnell GN, Allderdice PW, et al (1976) The 9p− syndrome. Ann Genet 19:11–16

André M-J, Aurias A, Berranger P, et al (1976) Trisomie 4p de novo par isochromosome 4p. Ann Genet 19:127–131

Berghe H van den, Fryns J-P, Cassiman J-J, et al (1974) Chromosome 6 en anneau. Caryotype 46,XY,r(6)/45,XY,−6. Ann Genet 17:29–35

Biederman B (1979) Trisomy 9p with an isochromosome of 9p. Hum Genet 46:125–126

Borgaonkar DS (1977) Chromosomal variation in man. A catalog of chromosomal variants and anomalies, 2nd edn. Liss, New York

Burnham CR (1962) Discussions in cytogenetics. Burgess, Minneapolis

Canki N, Dutrillaux B (1979) Two cases of familial paracentric inversion in man associated with sex chromosome anomaly. Hum Genet 47:261–268

Christensen KR, Friedrich U, Jacobsen P, et al (1970) Ring chromosome 18 in mother and daughter. J Ment Defic Res 14:49–67

Craig-Holmes AP (1977) C-band polymorphism in human populations. In: Hook EB, Porter IH (eds) Population cytogenetics. Academic, New York, pp 161–177

Dallapiccola B, Brinchi V, Curatolo P (1977) Variability of r(22) chromosome phenotypical expression. Acta Genet Med Gemellol 26:287–290

Francke U (1977) Abnormalities of chromosomes 11 and 20. In: Yunis JJ (ed) New chromosomal syndromes. Academic, New York, pp 245–272

Gorlin RJ (1977) Classical chromosome disorders. In: Yunis JJ (ed) New chromosomal syndromes. Academic, New York, pp 59–117

Grouchy J de, Turleau C (1977) Clinical atlas of human chromosomes. Wiley, New York

Hansmann I (1976) Structural variability of human chromosome 9 in relation to its evolution. Hum Genet 31:247–262

Hoo JJ, Obermann U, Cramer H (1974) The behavior of ring chromosome 13. Humangenetik 24:161–171

Hsu LYF, Hirschhorn K (1977) The trisomy 22 syndrome and the cat eye syndrome. In: Yunis JJ (ed). New chromosomal syndromes., Academic, New York, pp 339–368

Inouye T, Matsuda H, Shimura K (1979) A ring chromosome 9 in an infant with malformations. Hum Genet 50:231–235

Johnson VP, Mulder RD, Hosen R (1976) The Wolf-Hirschhorn (4p−) syndrome. Clin Genet 10:104–112

Kunze J, Stephan E, Tolksdorf M (1972) Ring-Chromosom 18-Ein 18p−/18q− Deletionssyndrom. Humangenetik 15:289–318

Kurnit DM (1979) Satellite DNA and heterochromatin variants: the case for unequal mitotic crossing over. Hum Genet 47:169–186

Lansky S, Daniel W, Fleiszar K (1977) Physical retardation associated with ring chromosome mosaicism: 46,XX,r(10)/45,XX,10−. J Med Genet 14:61–63

Leonard C, Hazael-Massieux P, Bocquet L (1975) Inversion péricentrique inv(2)(p11q13) dans des familles non apparentées. Humangenetik 28:121–128

Leschot NJ, Lim KS (1979) "Complete" trisomy 5p; de novo translocation t(2:5)(q36;p11) with isochromosome 5p. Hum Genet 46:271–278

Madan K, Bobrow M (1974) Structural variation in chromosome No. 9. Ann Genet 17:81–86

Magenis RE, Palmer CG, Wang L, et al (1977) Heritability of chromosome banding variants. In: Hook EB, Porter IH (eds) Population cytogenetics. Academic, New York, pp 179–188

Mayer M, Matsuura J, Jacobs P (1978) Inversions and other unusual heteromorphisms detected by C-banding. Hum Genet 45:43–50

Moorhead PS (1976) A closer look at chromosomal inversions. Am J Hum Genet 28:294–296

Nakajima S, Yanagisawa M, Kamoshita S, et al (1976) Mental retardation and congenital malformations associated with a ring chromosome 9. Hum Genet 32:289–293

Niebuhr E (1977) Partial trisomies and deletions of chromosome 13. In: Yunis JJ (ed) New chromosomal syndromes. Academic, New York, pp 273–299

Niebuhr E (1978a) Cytologic observations in 35 individuals with a 5p− karyotype. Hum Genet 42:143–156

Niebuhr E (1978b) The cri du chat syndrome. Hum Genet 44:227–275

Nielsen KB, Dyggve H, Friedrich U, et al (1978) Small metacentric nonsatellited extra chromosome. Hum Genet 44:59–69

Niss R, Passarge E (1975) Derivative chromosomal structures from a ring chromosome 4. Humangenetik 28:9–23

Ono K, Suzuki Y, Fujii I, et al (1974) A case of ring chromosome E17: 46,XX,r(17)(p13→q25). Jpn J Hum Genet 19:235–242

Opitz JM, Patau K (1975) A partial trisomy 5p syndrome. In: New chromosomal and malformation syndromes. Birth defects: original article series, Vol 11. The National Foundation-March of Dimes, New York, pp 191–200

Phillips RB (1978) Pericentric inversions inv(2)(p11q13) and inv(2)(p13q11) in 2 unrelated families. J Med Genet 15:388–390

Rethoré M-O (1977) Syndromes involving chromosomes 4, 9, and 12. In: Yunis JJ (ed) New chromosomal syndromes. Academic, New York, pp 119–183

Rodiere M, Donadio D, Emberger J-M, et al (1977) Isochromosomie 18: 46,XX,i(18q). Ann Pediat 24:611–616

Sanchez O, Yunis JJ (1977) New chromosome techniques and their medical applications. In: Yunis JJ (ed) New chromosomal syndromes. Academic, New York, pp 1–54

Schinzel A, Schmid W, Lüscher U, et al (1974) Structural aberrations of chromosome 18. I. The 18p– syndrome. Arch Genet 47:1–15

Schroeder H-J, Schleiermacher E, Schroeder TM, et al (1967) Zur klinischen Differentialdiagnose des Cri du Chat-Syndroms. Humangenetik 4:294–304

Shimba H, Ohtaki K, Tanabe K, et al (1976) Paracentric inversion of a human chromosome 7. Hum Genet 31:1–7

Smith DW (1976) Recognizable patterns of human malformation, 2nd edn. Saunders, Philadelphia

Soudek D, Sroka H (1978) Inversion of ''fluorescent'' segment in chromosome 3: a polymorphic trait. Hum Genet 44:109–115

Surana RB, Bailey JD, Conen PE (1971) A ring-4 chromosome in a patient with normal intelligence and short stature. J Med Genet 8:517–521

Tharapel AT, Summitt RL (1978) Minor chromosome variations and selected heteromorphisms in 200 unclassifiable mentally retarded patients and 200 normal controls. Hum Genet 41:121–130

Therkelsen AJ, Hultén M, Jonasson J, et al (1973) Presumptive direct insertion within chromosome 2 in man. Ann Hum Genet 36:367–373

Toomey KE, Mohandas T, Sparkes RS, et al (1978) Segregation of an insertional chromosome rearrangement in 3 generations. J Med Genet 15:382–387

Trunca Doyle C (1976) The cytogenetics of 90 patients with idiopathic mental retardation/malformation syndromes and of 90 normal subjects. Hum Genet 33:131–146

Trunca C, Opitz JM (1977) Pericentric inversion of chromosome 14 and the risk of partial duplication of 14q(14q31→14qter). Am J Med Genet 1:217–228

Uchida IA, McRae KN, Wang HC, et al (1965) Familial short arm deficiency of chromosome 18 concomitant with arhinencephaly and alopecia congenita. Am J Hum Genet 17:410–419

Vianna-Morgante AM, Nozaki MJ, Ortega CC, et al (1976) Partial monosomy and partial trisomy 18 in two offspring of carrier of pericentric inversion of chromosome 18. J Med Genet 13:366–370

Vine DT, Yarkoni S, Cohen MM (1976) Inversion homozygosity of chromosome no. 9 in a highly inbred kindred. Am J Hum Genet 28:203–207

Vogel W, Back E, Imm W (1978) Serial duplication of 10(q11→q22) in a patient with minor congenital malformations. Clin Genet 13:159–163

Wisniewski L, Politis GD, Higgins JV (1978) Partial tetrasomy 9 in a liveborn infant. Clin Genet 14:147–153

Wyandt HE, Magenis RE, Hecht F (1977) Abnormal chromosomes 14 and 15 in abortions, syndromes, and malignancy. In: Yunis JJ (ed) New chromosomal syndromes. Academic, New York, pp 301–338

Yunis E, Silva R, Egel H, et al (1978) Partial trisomy-5p. Hum Genet 43:231–237

Zachai EH, Breg WR (1973) Ring chromosome 7 with variable phenotypic expression. Cytogenet Cell Genet 12:40–48

XVI
Robertsonian Translocations

Occurrence

Robertsonian translocations refer to the recombination of whole chromosome arms. Such translocations take place most often between acrocentric or telocentric chromosomes. They have played an important role in the evolution of both plants and animals as demonstrated by organisms within a species (or in closely related species) that have different chromosome numbers but the same number of chromosome arms. In man, Robertsonian translocations are the most common structurally abnormal chromosomes. Surprisingly, they seem to be much rarer in some other animal species, such as mice or cattle (cf. Ford, 1970).

Whole-arm transfers between human chromosomes other than the acrocentrics seem to be extremely rare; only a couple of cases have been reported. One reason for this scarcity may be that such translocations are not ascertained through abnormal offspring, since monosomy for one chromosome arm combined with trisomy for another would be lethal (cf. Schober and Fonatsch, 1978). However, whole-arm transfers between nonacrocentric chromosomes are not found in unselected populations either. In the following discussion the term Robertsonian translocation will be used to describe a whole-arm transfer between acrocentrics.

Robertsonian translocations do not seem to affect the phenotype of a balanced carrier, apart from occasional male sterility. Individuals with 45 chromosomes including a Robertsonian translocation between two long arms are called balanced. That the deletion of the short arms of the acrocentrics does not have any damaging effects is a further indication of their total heterochromasy.

The relatively high frequency of Robertsonian translocations probably reflects their high incidence. On the other hand, the probability of their ascertainment is increased by their familial occurrence. For instance, 85 percent to 95 percent of DqDq translocations are familial in unselected material (cf. Nielsen and Rasmussen, 1976).

The different modes of formation of Robertsonian translocations between two acrocentric chromosomes are shown diagrammatically in Fig. VII.4. One chromosome may break through the short arm and the other through the long arm near the centromere, or both may break through the centromere. In both cases the result is one long and one short monocentric chromosome. If both chromosomes have a break in the short arm, one dicentric and one acentric chromosome are formed. The mechanics of the formation of Robertsonian translocations are discussed, for instance, by John and Freeman (1975).

Monocentric and Dicentric Chromosomes

Banding techniques demonstrate unequivocally that both monocentric and dicentric Robertsonian translocations exist (Niebuhr, 1972). However, in individual cases it is often difficult to distinguish between the two. This is not made easier by the fact that one centromere in a dicentric is often inactivated, which means that its site is not marked by a constriction (Niebuhr, 1972). However, even if both centromeres are functioning this does not lead to aberrations in mitosis, since there is almost never a twist between them.

The small chromosome consisting of the short arms of two acrocentrics is usually, but not invariably, lost. The occurrence of such small bisatellited chromosomes, in addition to the normal chromosome complement, was mentioned in the preceding chapter. Among 11,148 Danish newborn infants, six with such chromosomes were found, giving a frequency of 0.54/1000 (Nielsen and Rasmussen, 1975). Palmer et al (1969) have described a rare family with both a DqDq and a DpDp chromosome. Four family members had both, whereas two showed only the long chromosome.

Isochromosomes resulting from misdivision of the centromere cannot be distinguished from Robertsonian translocations between two homologous chromosomes on morphological grounds. However, the history of the chromosome often allows a determination to be made. If a person has one normal cell line and another with 46 chromosomes, including one free 21 and a metacentric consisting of two chromosomes 21, the latter is obviously an isochromosome. On the other hand, in a mosaic with a normal cell line and another with 45,t(21q21q), a Robertsonian translocation can be taken for granted. In a family studied in our laboratory, in which three children had Down's syndrome, the father was such a mosaic and the affected children had 46,t(21q21q).

The most probable explanation for the chromosome constitution of a girl mentioned previously (Therman et al, 1963), who had a mild 13 trisomy syndrome and two cell lines—one with tel(13q) and another with i(13q)—is misdivision of the centromere. Similar mosaicism involving chromosome 21 has also been found a few times (cf. Hornstein and Soukup, 1976).

Even more complicated mosaics involving Robertsonian translocations have been encountered. Thus in two children with features of Down's syndrome, one cell line had 45 chromosomes with a 15q21q translocation, whereas the other showed 46 chromosomes, including a 21q21q chromosome (Atkins and Bartsocas, 1974; Vianna-Morgante and Nunesmaia, 1978). A nonmosaic child with 44 chromosomes, including both a DqDq and a DqGq chromosome, was born to a couple in which the DqDq chromosome came from the mother's family and the DqGq from the father's (Orye and Delire, 1967). In another child with 13 trisomy syndrome, two DqDq translocations were found (Cohen et al, 1968).

Relative Frequencies of the Different Types of Robertsonian Translocations

Apart from 14q14q, all possible combinations of acrocentric chromosomes are found in Robertsonian translocations. It became clear at an early stage, however, that the participation and the combinations of the different chromosomes were highly nonrandom. Initially the identity of the chromosomes was determined with autoradiography, which in the D and G group works so well that when the same cases were later studied with banding techniques, practically all the previous determinations were verified.

In Table XVI.1 are the pooled observations of 376 Robertsonian translocations ascertained in various ways; the list is not complete. However, the relative frequencies would not be significantly different in a more comprehensive sample. Three error sources affect all such attempts: (1) The identification of the chromosomes is often unreliable in early studies. (2) The same case may be published more than once. (3) The rarer the chromosome constitution, the greater the likelihood that the case will be published, and consequently the rare combinations will be over-represented.

Of the DqDq translocations, by far the most common is the combination of 13 and 14. The same position among the DqGq translocations is occupied by 14q21q, whereas of the GqGq translocations 21q21q seems to be slightly more common than 21q22q. Although 14q14q has never been observed, the frequencies of combinations other than 13q14q and 14q21q are not very different from each other.

Table XVI.1. Frequencies of
Different Types of
Robertsonian Translocations
Ascertained in Various Ways.[a]

Chromosome

		13	14	15	21	22
	13	11				
Chromosome	14	129	0			
	15	4	8	5		
	21	10	129	19	27	
	22	5	4	5	15	5

[a]Data from 79 studies.

Naturally the mode of ascertainment of Robertsonian translocations has an effect on the figures obtained. Most of the published cases come from three sources: (1) unbiased studies, such as consecutive newborns or chance findings in populations selected for various reasons; (2) studies on families ascertained through an unbalanced individual; (3) studies on persons with infertility problems, including habitual abortions.

Newborn Studies

Hook and Hamerton (1977) have reviewed six studies in which the chromosome constitutions of 56,952 newborn infants were determined. In this material, 51 Robertsonian translocations were found of which 40 were DqDq and 11 were DqGq combinations. The incidence of Robertsonian translocations in an unselected population is thus 0.09 percent or about 1/1000 (Table XVII.2). In 21 of the DqDq translocations, the participating chromosomes have been identified; in 20 cases they were 13 and 14, and in one 14 and 15. Four of the identified DqGq translcations involved 14 and 21, one was 13q21q, one 14q22q, and one 15q22q (cf. Jacobs, 1977).

Ascertainment Through an Unbalanced Individual

A common mode of ascertainment for Robertsonian translocations is through an individual with Down's syndrome. In patients selected for Down's phenotype, the frequency of Robertsonian translocations ranges from 3.2 percent (Hongell et al, 1972) to 5 percent in 4330 patients (cf. Chapman et al, 1973). The mother's age is also a factor; about 8 percent of Down's syndrome patients whose mothers were under 30 years of age had translocations, whereas if the mothers were over 30, only 1.5 percent of the affected children were translocation cases (cf. Mikkelsen, 1971).

Mikkelsen (1971, 1973) lists 76 DqGq translocations from her laboratory and the literature, ascertained through a Down's syndrome patient. One of them had a t(13q21q), 65 had t(14q21q), 9 had 15q21q, and in one case the chromosome involved was either 14 or 15. Translocations between chromosomes 21 and 22 are often found through an affected person, and translocations between two chromosomes 21 are practically always found this way. Of 12 GqGq chromosomes ascertained in this way, 3 were 21q22q and 9 were 21q21q (Mikkelsen, 1973).

Translocations between chromosomes 13 and 14 are rarely discovered through a 13 trisomic individual, whereas this is the usual ascertainment for 13q13q translocations. Only occasionally has a DqGq chromosome been found through a 13 trisomic individual; one family with a t(13q21q) (Pérez-Castillo and Abrisqueta, 1978) and two families with 13q22q (Abe et al, 1975; Daniel and Lam-Po-Tang, 1976) have been detected in this way.

Ascertainment Through Infertility

Both sterility and habitual abortions are caused by certain types of Robertsonian translocations. This is to be expected in carriers of 15q15q translocations (Žižka et al, 1977) and in carriers of 22q22q translocations (cf. Farah et al, 1975; Mameli et al, 1978). Although some 22 trisomic children have been born alive, none of them has so far had a 22q22q translocation. The four pregnancies of a woman with 15q22q also ended in abortions (Fried et al, 1974).

The reproductive histories of families with a 13q14q translocation have as a rule been perfectly normal. However, some exceptions have been described, although it is often difficult to judge whether the reproductive troubles reported were in fact caused by the translocation or were incidental to it (cf. von Koskull and Aula, 1974).

Males with a DqDq translocation, usually between 13 and 14, have repeatedly been found to suffer from decreased fertility, often caused by oligospermia, although this is by no means a general feature in such carriers. Thus among 233 males with oligospermia, 8 DqDq translocation carriers were found, which amounts to 3.4 percent as compared with 0.1 percent in newborns (cf. Nielsen and Rasmussen, 1976). The reason for the occasional infertility of DqDq translocation carriers is not known.

Nonrandomness of Robertsonian Translocations

Several hypotheses have been put forward to explain the relatively high incidence of Robertsonian translocations and the highly nonrandom participation of the different acrocentrics in them (cf. Hecht and Kimberling, 1971; Mikkelsen, 1973). One assumption is that centric heterochromatin and the satellite stalks are especially prone to breakage. Another phenomenon characteristic of acrocentric chromosomes is their participation in satellite associations (Fig. VI.1 a and e), which may play an important role in bringing their centric regions together. It is of interest in this connection that in mouse cell lines the nucleolar chromosomes are preferentially involved in Robertsonian translocations (Miller et al, 1978).

Robertsonian translocations display features that distinguish them from other reciprocal translocations: (1) Ionizing radiation and chromosome-breaking substances do not seem to increase the incidence of Robertsonian translocations. On the other hand, certain substances, such as mitomycin C, preferentially cause these whole-arm transfers (Hsu et al, 1978). (2) Unlike other translocations, which take place more or less randomly between different chromosomes, the participation of the acrocentrics in Robertsonian translocations is extremely nonrandom (Table XVI.1). Why, when translocations between chromosomes 13 and 14 and between 14 and 21 are relatively common, should the combinations 13q13q and 13q21q be so rare and 14q14q nonexistent?

Most of the differences between Robertsonian and other translocations can be explained on the assumption that the former, as a rule, are the result of an exchange in meiosis or mitosis, and not of breakage and rejoining. Possibly the repeated sequences in the centric regions have an indiscriminate tendency to pair, and crossing-over in a reversely paired segment or a U-type exchange would lead to whole-arm transfers. This would be the mechanism producing the baseline number of Robertsonian translocations in the different classes (Table XVI.1). The behavior of chromosomes 13, 14, and 21, on the other hand, is understandable if we assume that they have in common a homologous segment (A–B in Fig. XVI.1), which is inverted in chromosome 14 relative to the two others. Crossing-over in the inverted region between chromosomes 13 and 14 and chromosomes 14 and 21, respectively, would give rise to t(13q14q) and t(14q21q), which are the common types of Robertsonian translocations.

A U-type exchange, instead of crossing-over, is the most plausible explanation for the formation of reverse tandem associations between two chromosomes 21 with one centromere inactivated and satellites on both ends (for example, Bartsch-Sandhoff and Schade, 1973; Schuh et al, 1974).

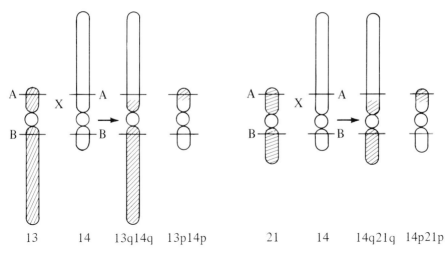

Fig. XVI.1. Origin of Robertsonian translocations 13q14q and 14q21q through crossing-over in segment A–B that is inverted in chromosome 14 relative to chromosomes 13 and 21.

Segregation in Carriers of Robertsonian Translocations

The existing information on segregation ratios in DqDq and DqGq carriers is summarized in Table XVI.2. All the offspring of a 13q13q and 21q21q carrier are 13 trisomic and 21 trisomic, respectively. Even a family with eight children with Down's syndrome in which the mother was a 21q21q carrier has been described (Furbetta et al, 1973). It may be stressed that for no other chromosomal defect than Robertsonian translocations between these homologous chromosomes is the risk of abnormal offspring 100 percent. The zygotes that are monosomic for either of these chromosomes end up as spontaneous abortions. Carriers of 15q15q and 22q22q—as might be expected—have no live offspring but only spontaneous abortions.

The risk figures for producing unbalanced offspring are somewhat different, depending on the mode of ascertainment. Whatever the ascertainment bias, the probability of producing a healthy carrier is 50 percent for 13q14q, 14q21q, and 21q22q carriers (Table XVI.2). In families found in unselected populations, such as newborn infants, the risk for unbalanced offspring is less than in families ascertained through an affected member. There is general agreement with the statement by Evans et al (1978, p. 112); "In conclusion, this and other similar studies suggest that when ascertained in a family by chance, both balanced reciprocal and Robertsonian, except t(14q21q), translocations carry a low risk of producing congenitally malformed offspring."

Table XVI.2.   Segregation in Carriers of Robertsonian Translocations.[a]

| Translocation | Risk of Unbalanced Offspring (%) | | Probability of Healthy Carrier (%) |
	Female carrier	Male carrier	
13qDq (mostly 14)	< 1	Very low	50
13q13q	100	100	0
Dq21q (mostly 14)	10	Very low, estimated 2.4	50
21q22q	6.8	Upper limit 2.9	50
21q21q	100	— 100	0

[a]Data from Hamerton, 1970; Mikkelsen, 1971; and Chapman et al, 1973.

For any carrier of a 13q14q translocation the risk is very low. Hamerton (1970) estimated the risk for a 13qDq carrier to be 0.67 percent, which is usually quoted as less than 1 percent. Low as this figure is, it is still higher than the risk for the general population, in which the estimates of the incidence of 13 trisomic children range from 0.005 percent to 0.02 percent. It should also be remembered that this risk figure (<1 percent) is based on one 13 trisomic offspring, who was not a proband (Hamerton, 1970).

Surprisingly, the probability of a DqDq translocation carrier producing a 21 trisomic child may be higher, the estimate being about 2 percent (Mikkelsen, 1971). However, more data are needed. Further, a number of cases have been reported in which a DqDq translocation is combined with other chromosome anomalies, such as XO, XXY, and XXYY. However, the observations are incidental, and for the present it is not known whether there is a causal relationship or whether these cases are simply a result of chance (Harris et al, 1979). A generally neglected factor is that the parental ages seem to be increased in these families (Schroeder, unpublished).

Most information on segregation of Robertsonian translocations comes from families ascertained through an individual with Down's syndrome (Table XVI.2). The risk figure for a female Dq21q carrier is about 10 percent. The risk for a male carrier is much lower. It has been estimated by Hamerton (1971) to be 2.4 percent, but this figure is based on a very small number of cases. The latest risk figure, 6.8 percent, for a female carrier of a t(21q22q) has been determined by Chapman et al (1973). For a male carrier the estimated upper limit for the risk is 2.9 percent.

The considerable difference in the risk of 13qDq and Dq21q carriers producing unbalanced offspring is assumed to depend on an alternate meiotic configuration being symmetrical for the former and asymmetrical for the latter (Fig. XVI.2). Segregation in the configuration formed by a DqDq translocation would, as a rule, result in one cell with the translocation chromosome and another with the two free chromosomes. In the

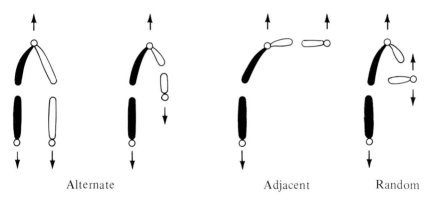

Alternate   Adjacent   Random

Fig. XVI.2. Different modes of meiotic orientation and segregation of Robertsonian translocations. Alternate segregation of t(DqDq) or t(DqGq) results in normals and carriers (1:1); adjacent segregation of t(DqGq) in trisomy and monosomy of G (1:1), and random drifting of univalent G in all four types (1:1:1:1).

meiotic configuration formed by Dq21q, the free 21 would be near the same pole as the translocation and would therefore sometimes undergo adjacent segregation. Probably also because 21 is smaller than a D chromosome, it would fail more often to form a chiasma and drift at random in the first meiotic anaphase. This may explain the observation that, although the alternate meiotic configuration formed by 21q22q is symmetrical, the risk for unbalanced offspring is much higher than for a DqDq carrier. Another factor that may influence the empirical risk figures is the possibility that more 13 trisomic than 21 trisomic zygotes end up as abortions.

Apart from 14q14q, all possible combinations of acrocentric chromosomes are found in Robertsonian translocations. However, the occurrence of the different combinations is highly nonrandom, the overwhelmingly most common being 13q14q and 14q21q. The differences between Robertsonian and other reciprocal translocations can be explained on the assumption that the former do not result from chromosomal breakage and rejoining, but rather from inverted pairing and crossing-over or from U-type exchanges. The highly nonrandom participation of 13, 14, and 21 in Robertsonian translocations is explicable if all three have a "homologous" region, which is in an inverted position on chromosome 14. The risk figures for various types of Robertsonian translocations are given in Table XVI.2.

References

Abe T, Morita M, Kawai K, et al (1975) Transmission of a t(13q22q) chromosome observed in three generations with segregation of the translocation D₁-trisomy syndrome. Humangenetik 30:207–215

Atkins L, Bartsocas CS (1974) Down's syndrome associated with two Robertsonian translocations, 45,XX,−15,−21,+t(15q21q) and 46,XX,−21,+t(21q21q). J Med Genet 11:306–309

Bartsch-Sandhoff M, Schade H (1973) Zwei subterminale Heterochromatinregionen bei einer seltenen Form einer 21/21-Translokation. Humangenetik 18:329–336

Chapman CJ, Gardner RJM, Veale AMO (1973) Segregation analysis of a large t(21q22q) family. J Med Genet 10:362–366

Cohen MM, Takagi N, Harrod EK (1968) Trisomy D₁ with two D/D translocation chromosomes. Am J Dis Child 115:185–190

Daniel A, Lam-Po-Tang PRLC (1976) Structure and inheritance of some heterozygous Robertsonian translocations in man. J Med Genet 13:381–388

Evans JA, Canning N, Hunter AGW, et al (1978) A cytogenetic survey of 14,069 newborn infants. III. An analysis of the significance and cytologic behavior of the Robertsonian and reciprocal translocations. Cytogenet Cell Genet 20:96–123

Farah LMS, de S Nazareth HR, Dolnikoff M, et al (1975) Balanced homologous translocation t(22q22q) in a phenotypically normal woman with repeated spontaneous abortions. Humangenetik 28:357–360

Ford CE (1970) The population cytogenetics of other mammalian species. In: Jacobs PA, Price WH, Law P (eds) Human population cytogenetics. Williams and Wilkins, Baltimore, pp 221–239

Fried K, Bukovsky J, Rosenblatt M, et al (1974) Familial translocation 15/22. A possible cause for abortions in female carriers. J Med Genet 11:280–282

Furbetta M, Falorni A, Antignani P, et al (1973) Sibship (21q21q) translocation Down's syndrome with maternal transmission. J Med Genet 10:371–375

Hamerton JL (1970) Robertsonian translocations. In: Jacobs PA, Price WH, Law P (eds) Human population cytogenetics. Williams and Wilkins, Baltimore, pp 63–80

Hamerton JL (1971) Human cytogenetics I. Academic, New York

Harris DJ, Hankins L, Begleiter ML (1979) Reproductive risk of t(13q14q) carriers. Case report and review. Am J Med Genet 3:175–181

Hecht F, Kimberling WJ (1971) Patterns of D chromosome involvement in human (DqDq) and (DqGq) Robertsonian rearrangements. Am J Hum Genet 23:361–367

Hongell K, Gripenberg U, Iivanainen M (1972) Down's syndrome. Incidence of translocations in Finland. Hum Hered 22:7–14

Hook EB, Hamerton JL (1977) The frequency of chromosome abnormalities detected in consecutive newborn studies—differences between studies—results by sex and by severity of phenotypic involvement. In: Hook EB, Porter IH (eds) Population cytogenetics. Academic, New York, pp 63–79

Hornstein L, Soukup S (1976) A case of atypical Down's syndrome with mosaic 46,XX/46,XX−21+t(21q21q). Clin Genet 10:77–81

Hsu TC, Pathak S, Basen BM, et al (1978) Induced Robertsonian fusions and tandem translocations in mammalian cell cultures. Cytogenet Cell Genet 21:86–98

Jacobs PA (1977) Structural rearrangements of the chromosomes in man. In: Hook EB, Porter IH (eds) Population cytogenetics. Academic, New York, pp 81–97

John B, Freeman M (1975) Causes and consequences of Robertsonian exchange. Chromosoma 52:123–136

Koskull H von, Aula P (1974) Inherited (13;14) translocation and reproduction. Humangenetik 24:85–91

Mameli M, Cardia S, Milia A, et al (1978) A further case of a 22;22 Robertsonian translocation associated with recurrent abortions. Hum Genet 41:359–361

Mikkelsen M (1971) Down's syndrome. Current stage of cytogenetic research. Humangenetik 12:1–28

Mikkelsen M (1973) Non-random involvement of acrocentric chromosomes in human Robertsonian translocations. In: Wahrman J, Lewis KR (eds) Chromosomes today, Vol 4. Wiley, New York, pp 253–259

Miller OJ, Miller DA, Tantravahi R, et al (1978) Nucleolus organizer activity and the origin of Robertsonian translocations. Cytogenet Cell Genet 20:40–50

Niebuhr E (1972) Dicentric and monocentric Robertsonian translocations in man. Humangenetik 16:217–226

Nielsen J, Rasmussen K (1975) Extra marker chromosome in newborn children. Hereditas 81:221–224

Nielsen J, Rasmussen K (1976) Autosomal reciprocal translocations and 13/14 translocations: a population study. Clin Genet 10:161–177

Orye E, Delire C (1967) Familial D/D and D/G₁ translocation. Helv Paediatr Acta 22:36–40

Palmer CG, Conneally PM, Christian JC (1969) Translocations of D chromosomes in two families t(13q14q) and t(13q14q)+(13p14p). J Med Genet 6:166–173

Pérez-Castillo A, Abrisqueta JA (1978) Patau's syndrome and 13q21q translocation. Hum Genet 42:327–331

Schober AM, Fonatsch C (1978) Balanced reciprocal whole-arm translocation t(1:19) in three generations. Hum Genet 42:349–352

Schuh BE, Korf BR, Salwen MJ (1974) A 21/21 tandem translocation with satellites on both long and short arms. J Med Genet 11:297–299

Therman E, Patau K, DeMars RI, et al (1963) Iso/telo-D₁ mosaicism in a child with an incomplete D₁ trisomy syndrome. Port Acta Biol 7:211–224

Vianna-Morgante AM, Nunesmaia HG (1978) Dissociation as probable origin of mosaic 45,XY,t(15:21)/46,XY,i(21q). J Med Genet 15:305–310

Žižka J, Balíček P, Finková A (1977) Translocation D/D involving two homologous chromosomes of the pair 15. Hum Genet 36:123–125

XVII
Reciprocal Translocations

Occurrence

Reciprocal translocations or interchanges have been observed in most organisms, plants as well as animals, which have been studied cytogenetically (cf. Burnham, 1962; White, 1978). They may occur as floating or stable polymorphisms or in single individuals. Just like other chromosome structural changes, such as Robertsonian translocations or inversions, translocations that are originally heterozygous may become homozygous, and this sometimes provides a mechanism to isolate two populations. The spontaneous rate of interchanges is estimated to lie between 10^{-4} and 10^{-3} per gamete per generation in such different organisms as *Drosophila,* grasshopper, mouse, and man (Lande, 1979).

In plants, balanced carriers of heterozygous translocations are usually discovered because a proportion of the pollen grains is abnormal in shape and size. In animals, balanced carriers are often detected because they are semisterile, and their litter sizes are smaller than normal. In man, the ascertainment is either through a phenotypically abnormal offspring or by chance in cytogenetic surveys of variously chosen populations.

Dicentric Human Chromosomes

When human reciprocal translocations are discussed, the other possible products (dicentric and acentric chromosomes) are often neglected, since it has been assumed that a dicentric chromosome is not viable, and the

		Number of Cases[b]	
Chromosome Arm		Partial trisomy	Partial monosomy
18	p	> 5	> 80
	q	> 10	> 50
19	p	—	—
	q	—	—
20	p	10	—
	q	—	—
21	q	> 10	8
22	q	23	—

[a]Data from de Grouchy and Turleau, 1977; Yunis, 1977; and Niebuhr, 1978.

[b]For most of the other arms individual cases of partial trisomy or monosomy are described.

in the Q-dark regions. However, when Aurias et al (1978) used several banding techniques on the same translocations, an excess of breaks was found at the borders of light and dark bands.

When balanced translocations have been ascertained by chance, the breakpoints seem to be more or less at random according to chromosome length (Jacobs et al, 1974; Jacobs, 1977). Of 95 reciprocal translocations in the study of Aurias et al (1978), 43 were detected through a balanced carrier. In the pooled material, an excess of breakpoints existed for chromosome arms 4p, 9p, 10q, 21q, and 22q, and a deficiency for 1p, 2p, and 16q. When the translocation was discovered through an unbalanced individual, there was an excess of breaks in the short arms and at telomeres. An interesting finding, although ascertainment bias may be partly responsible, is that among 21-trisomics and their parents there is an increased number of reciprocal translocations between unrelated chromosomes.

So far the largest study, unfortunately on pooled material that included all types of structurally abnormal chromosomes, has been done by Yu et al (1978). They analyzed the locations of 1134 breakpoints and found them to be nonrandom, with the longer chromosomes showing relatively fewer breaks than the shorter ones. The most breaks per unit length were found in chromosomes 9, 13, 18, 21, 22, and the Y, whereas chromosomes 2, 3, 6, 16 and 19 had the least.

Obviously answers to the questions about how nonrandomly the breakpoints in human translocations are situated and whether certain bands constitute special hot spots have to await a more extensive study in which the material is divided strictly according to ascertainment. Such a study is indeed under way (Trunca, unpublished).

Multiple Rearrangements

In many organisms, including man, individuals with complicated chromosome rearrangements have been found. Thus Palmer et al (1976) described in the mother of a chromosomally unbalanced child a rearrangement involving chromosomes 3, 11, and 20. They also listed 14 previously published multiple-break cases. Bijlsma et al (1978), who described an exceptional family with two reciprocal translocations t(2;6) and t(7;12) in three generations, leading to the birth of one unbalanced individual, mention a few more complex rearrangements.

A phenotypically abnormal child with five structurally aberrant chromosomes 1, 7, 4, 15, and 12 was born to a woman who, during pregnancy, developed malignant melanoma, which was not treated before the child was born (Fitzgerald et al, 1977). One cannot help wondering whether the same unknown agent might have been responsible for the malignant disease in the mother and the chromosome aberrations in the child. The record in multiple breaks may have been reached by a boy with de novo chromosome abnormalities involving six chromosomes. His mental retardation and minor anomalies seem to be caused by a small interstitial deletion in 10p (Prieto et al, 1978).

Based on present evidence, it is impossible to decide whether such complex chromosome rearrangements, although rare, might still be more common than expected if breaks occurred independently. Obviously publication bias is a significant factor, since the more complicated a chromosome rearrangement is, the more certain it is to be published. However, it is also true that many agents, such as cosmic rays, viruses, or mutagenic substances affect cells nonrandomly.

Phenotypes of Balanced Translocation Carriers

There is no doubt that the overwhelming majority of persons with a balanced reciprocal translocation are phenotypically normal. However, some observations indicate that at least certain reciprocal translocations may affect the phenotype. This evidence comes from studies in which the frequency of apparently balanced reciprocal translocations is compared in mentally retarded and control populations. For instance, Funderburk et al (1977) found in 455 retarded children seven reciprocal translocations, whereas the corresponding number in 1679 nonretarded children with psychiatric problems was four (P < 0.05). By pooling the results of various surveys on mental retardates, the authors conclude that these patients had five times more balanced, mainly de novo, reciprocal translocations than consecutive newborns. Interestingly, the incidence of balanced Robertsonian translocations does not seem to be increased in mentally retarded patients. Whether breaks in particular chromosome regions are more likely to have phenotypic effects, as suggested by Biederman and Bowen (1976), for 12q15-q21 is not clear for the present. Although rare, the possibility of phenotypic effects should not be neglected in genetic counseling when amniocentesis reveals an apparently balanced reciprocal translocation.

Numerous hypotheses have been put forward to explain the phenotypic effects of some apparently balanced reciprocal translocations (cf. Hecht et al, 1978). One of the most attractive is the assumption that a break may occur within a gene, thus destroying its function. Another possibility is a small deletion or duplication at the break site. Still another hypothesis is position effect, which means that a gene functions differently at a new location. Position effects have been suggested in many plants and animals, although very few proven cases exist.

It is clear that there are different kinds of position effects. The best known is the so-called variegation type, which has been studied in *Drosophila* and the mouse. A variegated phenotype is caused by a gene being transferred next to heterochromatin, which results in its being active in some cells but not in others. This kind of position effect has not been observed in man. Indeed the only definite evidence for position effect in human chromosomes is the observation that a certain region on Xq has to be intact in both X chromosomes to allow normal female sex development (cf. Therman et al, 1980).

It was already mentioned that balanced Robertsonian translocations seem to cause infertility in some male carriers. The frequency of apparently balanced reciprocal translocations is also increased in male patients of subfertility clinics. Oligospermia in them seems to be the result of arrested or impaired spermatogenesis (cf. Chandley et al, 1975). It has been suggested that this phenomenon too may be caused by a position effect. In male mice translocation carriers also range from fertile to those in whom the spermatogenesis has stopped at the onset of meiosis (Searle et al, 1978).

Phenotypes of Unbalanced Translocation Carriers

An unbalanced translocation involves partial trisomy for one chromosome and partial monosomy for another, although the deleted segment may be very small or almost nonexistent if one of the breaks lies at a telomere. Trisomy and monosomy effects practically always include mental retardation and multiple congenital anomalies. The analysis of the symptoms is often made difficult by simultaneous monosomy and trisomy. By now the chromosomes and phenotypes of hundreds, possibly thousands, of unbalanced translocation carriers have been studied. The reader is referred to the compendiums of Borgaonkar (1977) and de Grouchy and Turleau (1977) and to the numerous reviews in *New Chromosomal Syndromes* (1977).

Even if breaks occurred exclusively in Q-dark bands, the number of random interchanges would be enormous. However, the range of combinations that allow the birth of a live child is much more limited, since most sizable deletions are lethal, as are many duplications. This strong selection has led to a relatively high incidence of certain partial trisomies and monosomies, whereas others are extremely rare or nonexistent.

For the more frequent partial trisomies and monosomies, defined syndromes have been described. Table XVII.1 gives a somewhat simplified overview of those autosomal arms for which partial trisomy or monosomy syndromes have been established. Only syndromes for which the description is based on at least five cases are included, although for many of the other chromosome arms, individual cases of deletions and duplications have been found. The numbers in Table XVII.1 are taken from de Grouchy and Turleau (1977), *New Chromosomal Syndromes* (1977), and Niebuhr (1978). Since the appearance of these reviews, more cases have naturally been published. However, there is probably little change in the *relative* frequencies of the different types.

By far the most common of such unbalanced conditions is the cri du chat syndrome caused by a deletion of 5p. Chromosomes 13 and 18 are also often involved, as are 4p, 9p, and 22q. On the contrary, other chromosomes and arms, such as 2, 3, 6, 16, 17, 19, and 20q have rarely been found in unbalanced translocations. However, occasionally an infant is born alive whose chromosome constitution would usually result in embryonic death. Thus, for instance, an extremely miserable infant who lived only for 7 h was trisomic for most of 1q attached to 3p, which chromosome was a segregant of the mother's balanced translocation (1q−;3p+) (Norwood and Hoehn, 1974).

Even a crude estimate (Table XVII.1) shows how nonrandomly the partial trisomies and monosomies that occur with any appreciable frequency are distributed among chromosome arms. The following parameters probably determine this unequal distribution: (1) the length of the chromosome segment involved, (2) the Q-brightness of a segment (to which the number of cases seems to stand in a direct relationship); (3) individual genes that may act as trisomy or monosomy lethals; and (4) special "hot spots," short Q-dark bands, which have been assumed to contain a high density of genes, may also act as trisomy and monosomy lethals (Korenberg et al, 1978; Daniel et al, 1979).

Examples of Translocation Families

Only a few rather arbitrarily chosen translocation families will be presented here as examples. A typical family (Fig. XVII.1) ascertained through a chromosomally unbalanced proband was studied in our laboratory (unpublished). The abnormal daughter was severely mentally retarded and showed diverse congenital anomalies. Chromosome analysis revealed a too long 4q, whereas the mother and sister were balanced carriers of t(11q−;4q+) (Fig. XVII.1). The spontaneous abortions were not available for cytogenetic studies, but it is possible that they represented other types of unbalanced chromosome constitutions. Since the segment from 11q is apparently attached to the very end of 4q, the symptoms of the proposita are probably caused mainly by trisomy for about one-half of 11q. Partial trisomy for 11q corresponds to a more-or-less defined syndrome into which the proposita's phenotype also fits (cf. Francke, 1977).

The most common situation for a balanced translocation carrier is that no type of unbalanced offspring is viable. The next most common kind of translocation family is that in which only one type of unbalanced offspring—usually the one with the smallest deletion—is born alive. For instance, Centerwall et al (1976) have presented a four-generation pedigree in which segregation of a t(9p−;14q+) resulted in six affected individuals with partial trisomy 9p and four normal carriers.

Fortunately it is rare that more than one type of unbalanced offspring is born alive in the same translocation family. The occurrence of cri du chat syndrome and its trisomic "countertype" is discussed in a previous chapter. For a number of other translocations, a similar segregation has been described. To take a random example, Jacobsen et al (1973) analyzed three generations of a family in which 14 normal carriers had t(11q−;21q+), two abnormal segregants showed partial monosomy for 11q, and one had trisomy for the same segment. As might be expected, the monosomic individuals were more severely affected than was the trisomic one. A more typical result of a reciprocal translocation 7q−;21q+ in a male was the birth of an abnormal daughter trisomic for the end region of 7q and a spontaneous abortion whose chromosome displayed the corresponding deletion (Bass et al, 1973).

A reciprocal translocation in one spouse is usually not a significant cause for repeated abortions, as demonstrated by several cytogenetic studies on series of couples with such reproductive histories. However, in individual families, a reciprocal translocation may be the cause of habitual abortions. For instance, Nuzzo et al (1973) found a maternal translocation in which the entire 2q was attached to 1p, resulting in four recognized abortions and no livebirths. Another large translocation 3q−;4p+ in the father apparently resulted in sterility after the birth of a normal carrier son (Sarto and Therman, 1976).

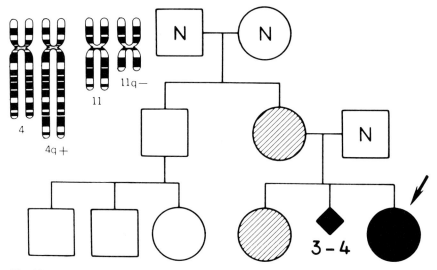

Fig. XVII.1. A family in which the mother and one daughter (shaded) were balanced carriers of t(4q+;11q−), one daughter (black) had the unbalanced translocation chromosome 4q+, and abortions possibly had 11q− in unbalanced state; N chromosomes normal.

A 1:3 segregation (one of the four chromosomes goes to one pole, while three go to the other) in a translocation carrier gives rise to unbalanced offspring with 45 or 47 chromosomes. Such disjunction is discussed in more detail in connection with the risk of abnormal offspring in translocation carriers. Sixteen kinds of offspring are possible from such a segregation. They include monosomy or trisomy for a normal chromosome. An extra translocated chromosome combined with an otherwise normal complement (47 chromosomes) has been termed *tertiary trisomy,* and a 45-chromosome complement that includes an abnormal chromosome is *tertiary monosomy.*

Interchange trisomy implies that the two translocated chromosomes are present together with an extra normal chromosome (47 chromosomes) (cf. Lindenbaum and Bobrow, 1975). A rare situation in which two different types of abnormal offspring resulted from 1:3 segregation is represented by a woman with t(9q−;21p+) who gave birth to one daughter with an extra 9p chromosome and another who was 21-trisomic, in addition to a chromosomally normal daugher (Habedank and Faust, 1978). Her translocation had features that are known to promote 1:3 disjunction, namely, the chromosomes were of different size, one was acrocentric, and the breaks were near the centromere.

In an interesting family with five abortions, three were studied cytologically by Kajii et al (1974). Segregation in the father who had a t(13q−;18q+) led to three chromosomally analyzed abortions; 47,+13 (tertiary trisomy), 46,13q− (unbalanced translocation), and 47,t(13q−;18q+)+18 (interchange trisomy). The abortion with 46 chromosomes resulted from an adjacent-1 2:2 disjunction (Fig. XVII.2), the other two from a 1:3 segregation.

Fig. XVII.2. Reciprocal translocation between two chromosomes, pachytene configuration of the four chromosomes, and modes of orientation of a ring of four in metaphase I: alternate, adjacent-1, adjacent-2 (1:3 segregation now shown). The gametes formed are: normal, balanced, two types of unbalanced (zygotes possibly viable), and two types of even more unbalanced (zygotes nonviable).

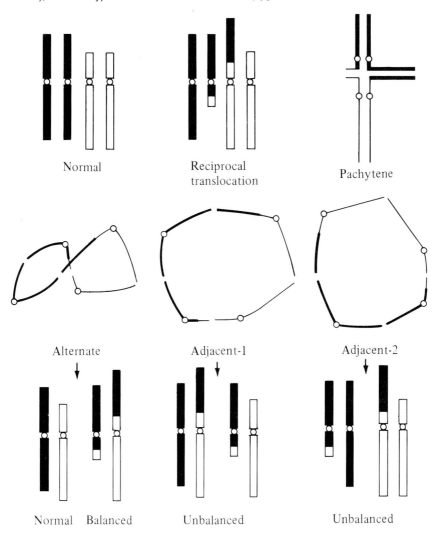

Normal   Reciprocal translocation   Pachytene

Alternate   Adjacent-1   Adjacent-2

Normal   Balanced   Unbalanced   Unbalanced

Meiosis in Translocation Carriers

Since homologous chromosome segments tend to pair in meiosis even when one of them is translocated to another chromosome, the two normal and the two translocation chromosomes often form a group of four, which in some organisms can be seen as a cross-like figure in pachytene (Fig. XVII.2). Very small translocated segments may remain unpaired or fail to form a chiasma, which results in the formation of two bivalents or a group of three and a univalent. However, if at least one terminal chiasma is present in each arm of the pachytene cross, the metaphase configuration will be a ring of four (Fig. XVII.2); or if one chiasma fails to form, the result will be a chain. A ring or chain of four may orient itself in different ways on the spindle.

An alternate orientation (Fig. XVII.2) gives rise to one cell with normal chromosomes and another with the translocation chromosomes. Adjacent-1 orientation (Fig. XVII.2) leads to the formation of two unbalanced cells, each with one translocation chromosome. Adjacent-2 orientation (Fig. XVII.2) gives rise to two cells that are even more unbalanced. Various types of 1:3 segregation are also possible. Sometimes two bivalents are formed, and their independent segregation produces equal numbers of normal, carrier, and two types of unbalanced gametes.

Although certain features in a translocation are known to promote 1:3 segregation (cf. Lindenbaum and Bobrow, 1975), there is *no theoretical expectation* for the segregation ratios of human reciprocal translocations, since the relative frequencies of the different orientations of a specific ring or chain are unknown. In plants and grasshoppers the following factors are reported to determine the relative frequencies of the different types of orientation (cf. Lewis and John, 1963): the relative lengths of the chromosomes and chromosome segments involved, the position of the centromeres and their distance from the breakpoints, the frequency and location of chiasmata, and premetaphase stretching.

Unfortunately almost no observations on the orientation of translocation configurations in human meiosis exist, so for the present we have little basis for predicting how a specific translocation will segregate. To provide a basis for more accurate predictions, the history of a large number of translocation families classified according to the chromosomes involved and the lengths of the exchanged segments, would be needed—a work which fortunately is in progress (Trunca, unpublished).

Genetic Risk for Translocation Carriers Ascertained by Chance

As in Robertsonian translocations, the empirical risk figures for a translocation carrier to produce abnormal offspring depend on the mode of ascertainment. One of the most important unbiased sources is provided by the newborn studies (cf. Hook and Hamerton, 1977). In 56,952 infants, 51 balanced reciprocal and insertional translocations were found, which happens to be the number of balanced Robertsonian translocations in the same material (Table XVII.2). This amounts to one balanced translocation per 1000 neonates. A much smaller number of unbalanced translocations or insertions, that is 7 (0.1/1000), was encountered in the same newborn population. Other findings of this study are given in Table XVII.2.

Family studies on randomly ascertained translocations show that such carriers have a very small risk of producing abnormal offspring, which is also the case in similarly ascertained carriers of Robertsonian translocations with the exception of t(14q21q) (cf. Jacobs et al, 1970, Jacobs et al, 1975).

It should be stressed that whatever the ascertainment of reciprocal translocations, the ratio of normal to balanced carrier offspring seems to be 1:1 (cf. Jacobs et al, 1970).

Genetic Risk for Reciprocal Translocation Carriers Ascertained Through an Unbalanced Individual

Cytogenetic surveys on variously chosen mental retardates show that chromosome abnormalities constitute an important cause of mental retardation and congenital anomalies. For instance in the Madison Blind Study, the chromosomes of 410 mentally retarded patients with congenital anomalies of unknown etiology were compared with the same number of normal controls. About 6 percent of the patients showed a structurally

Table XVII.2. Chromosome Abnormalities in 56,952 Newborn Infants.[a]

Abnormality	Incidence (per 1000 neonates)
Aneuploidy (including mosaicism)	
Sex chromosomes[b]	
Males	2.61
Females	1.51
Autosomes	1.44
Structural abnormalities	
Balanced Robertsonian translocations	0.90
Balanced translocations	0.90
Unbalanced Robertsonian translocations	0.07
Unbalanced translocations	0.12
Other structural abnormalities	0.55
Total	6.05

[a]Modified from Hook and Hamerton, 1977.

[b]Numbers apply to infants of the relevant sex.

abnormal, unbalanced chromosome constitution, whereas none of the controls did (Trunca, unpublished).

Jacobs et al (1978) summarized the results of a number of cytological studies on mentally retarded populations. If the patients were unselected, 2 percent had a chromosome anomaly, excluding those with Down's syndrome. In selected patients, which were generally more retarded, 11 percent showed an abnormal chromosome constitution. Including only the autosomal trisomies—again excluding 21-trisomy—and the the structurally unbalanced chromosome constitutions, the corresponding number is 8.4 percent.

The risk figures for carriers who have been ascertained through a chromosomally unbalanced family member are very different from those found by chance. The risk is also different for the female and the male carriers. Lejune et al (1970) estimated that the empirical risk figure for a female carrier detected through an affected individual is in the range of 10 percent to 20 percent. For a corresponding male carrier the risk is estimated to be one-half of this, in other words 5 percent to 10 percent.

Lindenbaum and Bobrow (1975) have reviewed the 1:3 segregation on the basis of 56 families in which it had occurred. They estimated that 6 percent to 15 percent of all chromosomally unbalanced translocation cases result from 1:3 disjunction. Sometimes 2:2 segregation leading to abnormal offspring takes place in the same carrier. As already mentioned, there are 16 different products of 1:3 disjunction; however, most of them have never been observed. The following factors seem to favor 1:3 over 2:2 segregation: one of the chromosomes is acrocentric, at least one break is near the centromere, and the participating chromosomes are very unequal in size. The translocations for which 1:3 segregation has been observed involve the chromosomes very nonrandomly. Thus interchange trisomy has only been found for chromosomes 18 and 21, whereas 9p seems to be preferred in tertiary trisomy (Lindenbaum and Bobrow, 1975).

About 90 percent of chromosomally unbalanced children resulting from 1:3 disjunction have been born to carrier mothers. The risks for further unbalanced liveborn offspring when the family has been ascertained through a 1:3 segregant are in the same range as for families that have been detected through an unbalanced 2:2 segregant, that is, 10 percent to 20 percent for a carrier mother and 0 for the father (Lindenbaum and Bobrow, 1975).

Interchromosomal Effects

Interchromosomal effects refer to the influence that a chromosome anomaly may have on the behavior of other chromosomes. For instance, a translocation or an inversion may affect the pairing, crossing-over, and segregation of unrelated chromosomes. Thus a DqDq translocation is claimed to increase the frequency of offspring with 21-trisomy or sex chromosome aneuploidy. Similarly, several families have been described in which a balanced reciprocal translocation is combined with monosomy or trisomy for unrelated chromosomes (cf. Aurias et al, 1978). However, whereas interchromosomal effects are a real possibility in man, the evidence is largely incidental and the selection of published cases is heavily influenced by both ascertainment and publication bias. The question about whether interchromosomal effects, indeed, exist in man or whether the described combinations of chromosome anomalies are to be ascribed to chance (or to increased parental age, Schroeder, unpublished), remains unsolved for the present.

References

Aurias A, Prieur M, Dutrillaux B, et al (1978) Systematic analysis of 95 reciprocal translocations of autosomes. Hum Genet 45:259–282

Bass HN, Crandall BF, Marcy SM (1973) Two different chromosome abnormalities resulting from a translocation carrier father. J Pediatr 83:1034–1038

Biederman B, Bowen P (1976) Balanced translocations involving chromosome 12: report of a case and possible evidence for position effect. Ann Genet 19:257–260

Bijlsma JB, France HF de, Bleeker-Wagemakers LM et al (1978) Double translocation t(7;12),t(2;6) heterozygosity in one family. Hum Genet 40:135–147

Borgaonkar DS (1977) Chromosomal variation in man. A catalog of chromosomal variants and anomalies, 2nd edn. Liss, New York

Burnham CR (1962) Discussions in cytogenetics. Burgess, Minneapolis

Centerwall WR, Miller KS, Reeves LM (1976) Familial "partial 9p" trisomy: six cases and four carriers in three generations. J Med Genet 13:57–61

Chandley AC, Edmond P, Christie S et al (1975) Cytogenetics and infertility in man. I. Karyotype and seminal analysis. Ann Hum Genet 39:231–254

Daniel A (1979) Structural differences in reciprocal translocations. Potential for a model of risk in Rcp. Hum Genet 51:171–182

Fitzgerald PH, Miethke P, Caseley RT (1977) Major karyotypic abnormality in a child born to a woman with untreated malignant melanoma. Clin Genet 12:155–161

Francke U (1977) Abnormalities of chromosomes 11 and 20. In: Yunis JJ (ed) New chromosomal syndromes. Academic, New York, pp 245–272

Funderburk SJ, Spence MA, Sparkes RS (1977) Mental retardation associated with "balanced" chromosome rearrangements. Am J Hum Genet 29:136–141

Grouchy J de, Turleau C (1977) Clinical atlas of human chromosomes. Wiley, New York

Habedank M, Faust J (1978) Trisomy 9p and unusual translocation mongolism in siblings due to different 3:1 segregations of maternal translocation rcp(9;21)(p11;q11). Hum Genet 42:251–256

Hecht F, Kaiser-McCaw B, Patil S, et al (1978) Are balanced translocations really balanced? Preliminary cytogenetic evidence for position effect in man. In: Summitt RL, Bergsma D (eds) Sex differentiation and chromosomal abnormalities. Birth defects: original article series, Vol 14, No. 6C. The National Foundation-March of Dimes, New York, pp 281–286

Hook EB, Hamerton JL (1977) The frequency of chromosome abnormalities detected in consecutive newborn studies—differences between studies—results by sex and by severity of phenotypic involvement. In: Hook EB, Porter IH (eds) Population cytogenetics. Academic, New York, pp 63–79

Jacobs PA (1977) Structural rearrangements of the chromosomes in man. In: Hook EB, Porter IH (eds) Population cytogenetics. Academic, New York, pp 81–97

Jacobs PA, Aitken J, Frackiewicz A, et al (1970) The inheritance of translocations in man: data from families ascertained through a balanced heterozygote. Ann Hum Genet 34:119–131

Jacobs PA, Buckton KE, Cunningham C et al (1974) An analysis of the break points of structural rearrangements in man. J Med Genet 11:50–64

Jacobs PA, Frackiewicz A, Law P, et al (1975) The effect of structural aberrations of the chromosomes on reproductive fitness in man. II. Results. Clin Genet 8:169–178

Jacobs PA, Matsuura JS, Mayer M, et al (1978) A cytogenetic survey of an institution for the mentally retarded: I. Chromosome abnormalities. Clin Genet 13:37–60

Jacobsen P, Hauge M, Henningsen K et al (1973) An (11;21) translocation in four generations with chromosome 11 abnormalities in the offspring. Hum Hered 23:568–585

Kajii T, Meylan J, Mikamo K (1974) Chromosome anomalies in three successive abortuses due to paternal translocation, t(13q–18q+). Cytogenet Cell Genet 13:426–436

Korenberg JR, Therman E, Denniston C (1978) Hot spots and functional organization of human chromosomes. Hum Genet 43:13–22

Lambert JC, Ferrari M, Bergondi C, et al (1979) 18q– syndrome resulting from a tdic(14p;18q). Hum Genet 48:61–66

Lande R (1979) Effective deme sizes during long-term evolution estimated from rates of chromosomal rearrangement. Evolution 33:234–251

Lejeune J, Dutrillaux B, Grouchy J de (1970) Reciprocal translocations in human populations. A preliminary analysis. In: Jacobs PA, Price WH, Law P (eds) Human population cytogenetics. Williams and Wilkins, Baltimore, pp 81–87

Lewis KR, John B (1963) Spontaneous interchange in *Chorthippus brunneus.* Chromosoma 14:618–637

Lindenbaum RH, Bobrow M (1975) Reciprocal translocations in man. 3:1 meiotic disjunction resulting in 47– or 45– chromosome offspring. J Med Genet 12:29–43

Niebuhr E (1978) The cri du chat syndrome—epidemiology, cytogenetics, and clinical features. Hum Genet 44:227–275

Norwood TH, Hoehn H (1974) Trisomy of the long arm of human chromosome 1. Humangenetik 25:79–82

Nuzzo F, Giorgi R, Zuffardi O, et al (1973) Translocation t(1p+;2q–) associated with recurrent abortion. Ann Genet 16:211–214

Palmer CG, Poland C, Reed T, et al (1976) Partial trisomy 11,46,XX,–3,–20,+der3,+der20,t(3:11:20), resulting from a complex maternal rearrangement of chromosomes 3, 11, 20. Hum Genet 31:219–225

Prieto F, Badia L, Moreno JA, et al (1978) 10p– syndrome associated with multiple chromosomal abnormalities. Hum Genet 45:229–235

Sarto GE, Therman E (1976) Large translocation t(3q–;4p+) as probable cause for semisterility. Fert Steril 27:784–788

Searle AG, Beechey CV, Evans EP (1978) Meiotic effects in chromosomally derived male sterility of mice. Ann Biol Anim Biochem Biophys 18:391–398

Therman E, Denniston C, Sarto GE, et al (1980) X chromosome constitution and the human female phenotype. Hum Genet 54: 133–143

White MJD (1978) Modes of Speciation. Freeman, San Francisco

Yu CW, Borgaonkar DS, Bolling DR (1978) Break points in human chromosomes. Hum Hered 28:210–225

Yunis JJ (ed) (1977) New chromosomal syndromes. Academic, New York

XVIII
Chromosomes and Cancer

What is Cancer?

There are two types of neoplasms, both of which are expressions of abnormal growth. Benign tumors are outgrowths that are self-limiting, that is, they grow to a certain size and then stop or regress. Most of us are acquainted with benign tumors, such as polyps or warts. Malignant tumors, on the other hand, usually show unlimited growth. They escape the rules of differentiation and grow wild. Malignant tumors also have the ability to infiltrate and destroy normal tissues. Most malignant tumors are also capable of spreading to new sites by metastases.

The histological structure of cancerous tumors appears to be disorganized and anaplastic and often differs greatly from the original normal tissue. An advanced cancer displays a confused cytological picture; cells with small and large, often giant, nuclei exist side by side with cells having several or weirdly shaped nuclei. Cancer mitoses also exhibit a wide spectrum of abnormalities, such as multiple poles, disorganized metaphases, anaphases with laggards and chromatid bridges, endomitoses, and endoreduplications. These phenomena indicate that the mitotic processes of replication and division, which are usually so orderly, have been dramatically disrupted.

Malignant tumors are by no means specific to man but are found throughout the animal kingdom from ants to whales. Corresponding phenomena also occur in plants. The so-called crown gall, for example, which affects a wide variety of plant species, resembles animal cancer. The induction of crown gall requires that a wound in a suitable host plant be inoculated with *Agrobacterium tumefaciens*. After enough cells have been transformed into crown gall cells, the tumor continues to grow even after the bacteria have been killed.

Agents Inducing Cancer

One of the confusing facts about cancer is that such a diversity of agents with nothing apparently in common has been found to induce malignant tumors. Thus exposure to all kinds of ionizing radiations and to ultraviolet light may cause this disease, and the number and variety of cancer-causing chemicals are almost infinite. Their genetic constitutions obviously play an important role in predisposing both mice and men to cancer. Viruses are known to induce malignant tumors in many animal species. Even mechanical irritation and exposure to heat are carcinogenic.

An interesting feature of cancer induction is that the disease often appears 10 to 30 years after the person has been exposed to a carcinogenic agent. Defects in the immunological defense mechanisms seem to play an important role in furthering tumor development and may even be the cause of the disease. That cancer is often a disease of old age may result from the weakening of immunological resistance in older persons.

Hypotheses on the Origin of Cancer

Cancer is caused by a mutation in the widest sense of the word. There is convincing evidence that the overwhelming majority of malignant tumors have a clonal origin (cf. Nowell, 1976); a change that takes place in a single cell is inherited by all its descendants. In primary tumors all cells may display the same abnormal chromosome constitution. Enzyme studies on women with different alleles of an X-linked gene show that the same X chromosome is active in all the cells of a tumor. The immunoglobulin produced by a plasma-cell tumor practically always confirms the assumption of a clonal origin of the disease.

The many facets of malignant growth have inspired an equally large number of hypotheses about its origin. (1) One or more gene mutations are assumed to be responsible. (2) A chromosome mutation that includes changes in the number or structure, or both, of chromosomes seems to be indicated at least in some cases. (3) Viruses are known to cause cancer in many animal species. (Whether they also play a general role in cancer induction in man is an open question.) (4) A mutation can take place in an extranuclear cellular organelle, for instance, a plasmid. In the plant tumor—crown gall—a segment of a plasmid is transferred from the inducing bacterium to the host cell and incorporated into it. (5) Malignant growth has been interpreted as a phenomenon of disturbed differentiation. However, this hypothesis has been more or less abandoned. (6) It has also been proposed that cancer mutations may be common in certain tissues and can develop into tumors in persons who have a weakness in the immunological resistance system.

Chromosomes and Cancer

It has been obvious since the beginning of the present century that the number of chromosomes in cancer cells often deviates greatly from the usual number in healthy cells of the host organism. The prominent German biologist Theodor Boveri observed that multipolar mitoses in sea urchin eggs led to abnormal chromosome numbers and these in turn led to abnormal development of the larvae. Since cancer cells often display multipolar divisions, Boveri concluded that the resulting deviant chromosome numbers were the cause of cancerous growth. Boveri's book *Zur Frage der Entstehung maligner Tumoren* ("On the problem of the origin of malignant tumors") appeared in 1914 (cf. Wolf, 1974). However, as has so often happened in cancer research, Boveri had put the cart before the horse. It is now clear that multipolar divisions appear only *after* the cells have undergone a malignant transformation. But Boveri's basic hypothesis, that chromosome aberrations may cause cancer, is very much alive today, although the final proof for it is still lacking.

In addition to this main problem of the relationship between chromosome aberrations and the origin of cancer, interesting questions are also posed by the chromosomal changes that a tumor undergoes during its development. What effects do such changes have on the growth and character of the tumor?

Cytological Techniques in Cancer Research

It has been assumed that the most obvious way to show that a specific chromosome change is the cause of a certain type of malignancy is to analyze the chromosomes of many tumors of the same type, looking for the same abnormal chromosome constitution in each one. The enormous amount of work done in this field has, however, proved to be frustrating. Many breakthroughs have been announced, which later turned out to be premature.

A truly staggering number of tumors, both primary and transplanted, has been analyzed cytologically. For example, tumors that grow as ascites fluid in the abdominal cavity of the host yield excellent chromosome preparations. Leukemic cells also lend themselves fairly well to chromosome studies, although the preparations—for unknown reasons—are rarely as satisfactory as those made from normal lymphocytes.

The study of solid tumors, on the other hand, presents serious technical difficulties. It is often impossible to make direct chromosome preparations that are of really high quality. Malignant tumors can be studied in tissue cultures, which usually yield satisfactory chromosome preparations. However, since conditions in vitro are different from those in vivo, selection may change the chromosome complement of the cells before analysis is possible.

Since the chromosome constitution of a malignant tumor often begins to change soon after transformation has taken place, it is important to be able to analyze its chromosomes at the very beginning of tumor development. However, this is often impossible. Primary tumors, especially in man, are difficult to detect, and it is sometimes impossible to decide whether or not a tumor is malignant at an early stage. This uncertainty is reflected, for instance, by the continuing discussion about whether carcinoma in situ and other similar conditions, in fact, are malignant.

Chromosome Studies in Ascites Tumors

The modern era of cancer cytology was launched in the 1950s with the study of transplantable ascites tumors of the mouse and the rat (cf. Yosida, 1975). They were also the first mammalian tissues from which satisfactory chromosome preparations were obtained. Bayreuther (1952) seems to have been the first investigator to apply a colchicine derivative to two mouse ascites tumors. He observed that both the chromosome number and morphology deviated from observations in normal mouse cells.

Subsequent studies have shown that many of the mouse ascites tumors are near-triploid or near-tetraploid and that the karyotype shows many morphologically abnormal chromosomes. Thus the hypotetraploid Ehrlich tumor, so widely used in various experiments, displays only a couple of chromosomes that can be matched, even approximately, with any normal mouse chromosome. Obviously the long-transplanted tumors are too far removed from the normal tissue of their origin to tell us anything about the origin of cancer.

Primary tumors in man sometimes induce the development of ascites fluid in which dividing tumor cells can be studied. Such cells also show striking abnormalities, both in chromosome number and structure.

Various strains of so-called HeLa cells are grown in tissue culture; they originated from a human cervical cancer. The HeLa lines are near-triploid and exhibit many structurally changed chromosomes (cf. Heneen, 1976).

Chromosome Studies in Primary Tumors

Despite the difficulties just pointed out, a great number of primary tumors and leukemic conditions have been analyzed cytologically. The results are reviewed in numerous articles and books (for example, Atkin,

1974; Sandberg and Sakurai, 1974; Makino, 1975; Müller and Stalder, 1976; Sandberg, 1979).

The recent review by Levan et al (1977) summarizes concisely the present knowledge of chromosomes and cancer: (1) Even some very malignant tumors have apparently normal chromosome complements. However, the existence of very small changes cannot be ruled out, even with best banding techniques.

(2) The majority of malignant tumors have chromosome complements that are abnormal, both in chromosome number and structure. Many tumors display so-called *marker chromosomes* that are morphologically abnormal.

(3) Most of the cells in a tumor belong to one *stemline,* which consists of cells with the same chromosome constitution, often including striking marker chromosomes. One or more sidelines with specific chromosome constitutions may also be present. A stemline has the ability to respond to a new environment, for instance after transplantation into an alien host species, by changing its chromosome constitution. (Some tumors have lost their immunological specificity to the extent that they are even able to grow in a different host species.) Chemotherapy may also affect the karyotype of a stemline (cf. Yosida, 1975). Finally the stemline usually changes during the development of a tumor.

(4) In their latest survey of 856 cases of human cancer in which at least one chromosome anomaly was found, Mitelman and Levan (1978) arranged, somewhat arbitrarily, the malignant diseases into 15 groups. The aberrations in them tended to cluster around specific chromosomes (those anomalies occurring in less than 20 percent of the cells were excluded). Of the 24 different human chromosomes only 12 participated in the anomalies. Chromosome 1 was involved in nine groups of diseases, chromosome 3 in three, chromosome 5 in three, chromosome 7 in four, chromosome 8 in eight, chromosome 9 in four, chromosome 13 in one, chromosome 14 in six, chromosome 17 in two, chromosome 20 in one, chromosome 21 in two, and chromosome 22 in four, whereas the rest of the chromosomes did not take part in the anomalies. It is obvious that the participation of the different chromosomes in the abnormalities characteristic of human cancer stemlines is extremely nonrandom (cf. also Rowley, 1977).

Specific Chromosome Abnormalities in Some Cancers

Among the confusing variety of chromosome aberrations in malignant cells, a few seem to stand out as being specific to a certain disease. The first of them was the so-called Philadelphia chromosome (Ph[1]) described by Nowell and Hungerford (1960). They reported that about one-half of the long arm of a G chromosome was missing in the great majority of patients suffering from chronic myeloid leukemia (CML). However, a minority of CML cases seemed to have normal chromosome complements. Based on the banding techniques, the deleted G chromosome was identified as a 22. A careful analysis by Rowley (1973) revealed that the segment missing from chromosome 22 was attached to the distal end of 9q. In a few cases the segment was translocated to other chromosomes, such as 2, 3, 4, 6, 11, 13, 16, 17, 21, and 22 (cf. Levan et al, 1977).

For a long time, CML appeared to be the only convincing example of a malignant disease being caused by a specific chromosome deletion. However, the finding that the aberration is in reality a reciprocal translocation makes the whole matter very puzzling, since such translocations do not as a rule have any phenotypic effects. The most reasonable explanation, at least for now, seems to be that we have here a position effect. In other words, chromosome 22 has to be continuous for the cell to function normally. Although in most cases the recipient chromosome is 9, other chromosomes are also involved. Therefore the identity of the other translocation chromosome appears to be of secondary importance. But that leaves us with the riddle: Why should chromosome 9 be involved so frequently in the translocation?

In the cells with the primary translocation between 22 and another chromosome, other chromosome aberrations appear during the development of CML (Fig. XVIII.1). Mitelman et al (1976) listed additional chromosome changes in 66 of 200 CML patients who had the Ph[1] chromosome. In 88 percent of the cases the chromosomal development took what the authors call the major route. In 18 patients, a second Ph[1] chromosome appeared, in seven an extra chromosome 8, whereas in nine patients an i(17q) was found. In nine patients both two Ph[1] chromosomes and an extra 8 occurred, whereas in six others the extra 8 was combined with an i(17q). Finally in six patients the cells showed, in addition to two Ph[1] chromosomes, both an extra 8 and i(17q). The same type of development is found when the disease is followed in individual patients (cf. de Grouchy and Turleau, 1974). Obviously the chromosomal evolution in CML is not random.

A problem similar to CML is posed by Burkitt's lymphoma. This is a human malignant disease that is exceptional in that a virus is assumed to be its cause. A reciprocal translocation between 8q and 14q is found repeatedly; a band from 8q is translocated to the distal end of 14q (Zech et al, 1976).

In human meningiomas, which are benign brain tumors, the cells first lose chromosome 22 and thereafter chromosome 8 (Mark, 1974). In other types of human tumors, such as breast and ovarian cancers, specific chromosomal changes are also reported.

Probably the most significant finding in man is that in many cases of

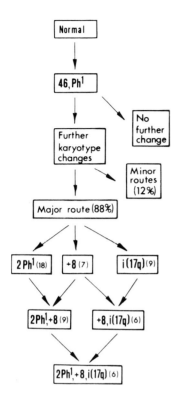

Fig. XVIII.1.   Chromosome abnormalities in addition to the Ph¹ chromosome in 66 cases of chronic myeloid leukemia. Figures in parenthesis in the lower part of the diagram indicate the number of cases showing the chromosome abnormality (Mitelman et al, 1976).

sporadic, unilateral retinoblastoma (cancer of the eye), the band 13q14 is missing (cf. Francke, 1976). In addition, most patients show moderate growth retardation and a characteristically similar facial appearance. The cause of bilateral retinoblastoma seems to be a dominant gene. The somatic cells of these patients do not reveal any chromosome abnormalities. However, it would be interesting to know if the tumor cells have undergone a deletion involving 13q14.

In dogs there is a type of malignant tumor that spreads like a venereal disease. The abnormal karyotype of the tumor cells seems to be the same in different dog varieties all over the world (cf. Makino, 1974). One possible explanation is that the same tumor line has spread throughout the world as an infectious disease.

Factors Promoting Tumor Development

That an organism's genetic constitution may have a decisive effect on its probability of developing cancer is borne out by many observations. For instance, there are inbred mouse strains in which most animals develop malignant tumors. A definite tendency to develop malignant disease seems to be inherited in some human families. Cancer Syndrome Families are described in which about one-half the members succumb to diseases of this group, suggesting that a dominant gene is the determining factor. In one such family (Lynch et al, 1977), 20 of 88 relatives studied had been affected with carcinoma of the colon and/or endometrium (uterine epithelium).

Other types of tumors were also found in the family, and 16 individuals had more than one primary malignancy. For offspring of affected parents the risk for colon/endometrial cancer was 52.8 percent in the 20- to 60-year age group, whereas none of the children of nonaffected parents or the unrelated spouses developed cancer. It would be of great interest to know whether chromosome aberrations are increased in the somatic cells of those bearing the "cancer gene."

Miller (1967) has reviewed the populations with specially high risk of leukemia. In Caucasian children under 15, the incidence of leukemia is 1/2880. If one identical twin has leukemia, the risk for the other is 1/5. In persons who have been exposed to ionizing radiation, the risk is greatly increased. Thus in Hiroshima survivors who were within 1000 m of the hypocenter, the probability of developing leukemia is 1/60.

Especially interesting is a group of diseases, each caused by a recessive gene, in which chromosomes break spontaneously. The most extensively studied of these are ataxia-telangiectasia, Fanconi's anemia, and Bloom's syndrome. The homozygotes have a greatly increased risk of malignant disease; for instance, in Bloom's syndrome it is 1/8. It is not known whether gene mutations also occur at a higher rate in the somatic cells of the affected individuals.

The conclusion seems inevitable that malignant disease depends to a great extent on genetic factors. It is also known that most, possibly all, carcinogens are mutagens; however, not all mutagens necessarily induce cancer. There is also a direct connection between the ability of various agents to break chromosomes and their carcinogenicity. Examples of such agents include ionizing radiation, a great variety of substances, diseases such as Bloom's syndrome, and viruses.

Although definite proof linking virus infections to human cancer is still lacking, many observations support the hypothesis. Viruses cause malignant tumors in many animal species. Of these animal viruses, SV40 breaks chromosomes in human tissue culture and also accomplishes a malignant transformation of the cells (cf. Harnden, 1974). Epidemiological studies link cervical cancer to infection of the female genital tract by herpes simplex virus (cf. Harnden, 1974). However, the most likely candidate for a virus-induced malignant disease in man is Burkitt's lymphoma, which is thought to be caused by the Barr-Epstein virus (cf. Harnden, 1974).

Many of these observations suggest the chromosomal origin of at least some malignancies. However, just when more and more evidence seemed to link chromosome aberrations to the origin of cancer, the Levans (father Albert and son Göran) and their co-workers in Sweden performed a series of relevant experiments that seriously undermine this hypothesis.

The Apparent Predetermination of Chromosome Changes in Cancer

The Levan group used two different agents to induce tumors, Rous sarcoma virus (RSV) and 7-12-dimethylbenz(α)anthracene (DMBA). In mice, rabbits, Chinese hamsters, and rats, comparable results were obtained (cf. Levan et al, 1977). Let us consider the results of the rat experiments. Primary sarcomas induced either by RSV or DMBA start out with the normal chromosome complement of the rat (Fig. XVIII.2). Gradually cells with abnormal chromosome constitutions appear and undergo a sequence of apparently predetermined chromosome changes, each agent inducing a specific nonrandom evolutionary pattern of its own. The origin of the tissue does not seem to affect the cytological development of the tumor.

In the RSV-induced tumors, the cells first gain an extra 7 chromosome, then a 13 chromosome, and finally a 12 chromosome (Fig. XVIII.2). The pattern of evolution in the DMBA-induced sarcomas, on the other hand, is different. First an extra 2 chromosome is added and then one of the small metacentric chromosomes. Similar experiments are discussed by DiPaolo and Popescu (1976).

The chromosome changes just described cannot possibly be the cause of the malignant tumors, because they appear only after the cells are transformed. The only reasonable explanation for these interesting results seems to be that each carcinogen induces a specific mutation that transforms the cell. This may be a gene mutation or a chromosome change too small to be discovered by the available techniques. The second step occurs when mitotic aberrations, which are characteristic of all malignant tumors, create numerous aberrant chromosome constitutions, and selection promotes the ones that divide the fastest. Each change in turn makes the cell more competitive and malignant.

Mitotic Aberrations in Cancer Cells

By the end of the last century it was known that multipolar mitoses are common in human cancer. Since then, evidence of a great variety of mitotic aberrations has accumulated. Chromosome breaks and structural

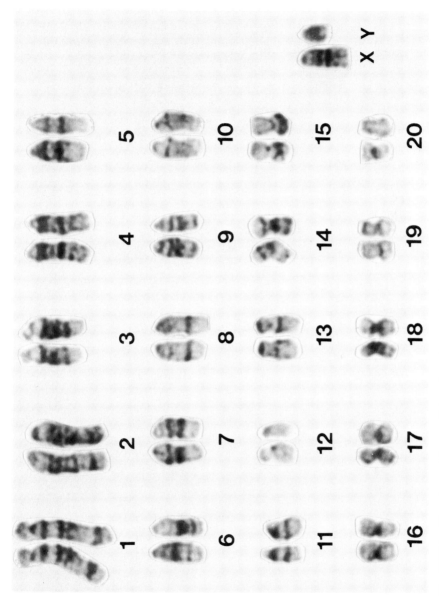

Fig. XVIII.2. Karyotype of the rat (G-banding) (courtesy of G Levan).

rearrangements are also much more frequent in cancer than in normal cells (cf. Shaw and Chen, 1974). The cytological picture in an advanced cancer is so confused that only by studying a whole range of tumors from the early stages to highly anaplastic ones, can one obtain an idea of the order in which the different aberrations appear and what their mutual relationships are. A series of such analyses on tumors ranging from early to highly malignant was performed around 1950 by Oksala, Therman, and Timonen (cf. Oksala and Therman, 1974) on Feulgen squash preparations of primary human cancers of the female genital tract.

In normal human tissues, prophase and metaphase require about the same time; this fact is reflected in their equal frequencies in counts on fixed biopsies. The ratio of metaphases to prophases is approximately 1 in normal tissues. In malignant tumors the ratio is increased and becomes higher in relation to the growing malignancy of the tumor, finally showing values of 23 to 35 (Fig. XVIII.3) (cf. Oksala and Therman, 1974).

When the metaphase/prophase ratio reaches values of 4 to 6, the first multipolar divisions appear. These are always tripolar. The incidence of tripolar mitoses goes up with an increasing ratio, and divisions with higher numbers of poles begin to appear (Therman and Timonen, 1950). In addition to these basic phenomena, cancer cells display an almost infinite variety of other mitotic aberrations, such as endoreduplication, endomitosis, C-mitosis, lagging chromosomes and chromatid bridges, as well as restitution at various stages (Chapter IV). Some of them may be

Fig. XVIII.3. Relative frequencies of prophases (P), metaphases (M), and anaphases (A) in three biopsies of normal epithelium of human fetal tubes (left) and three cases of cancer in adult fallopian tubes (data from Lehto, 1963).

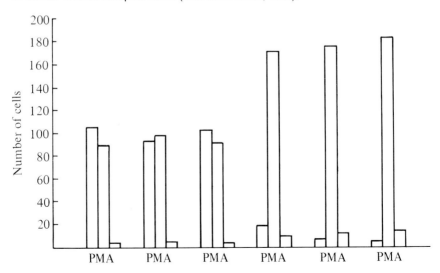

secondary results of abnormal tumor physiology, especially anoxia (lack of oxygen). These results have been confirmed repeatedly in different types of human cancers, as well as transplantable and induced mouse tumors (cf. Oksala and Therman, 1974).

The mitotic aberrations in malignant cells, especially the increased metaphase/prophase ratio and the occurrence of multipolar divisions, are so characteristic of cancer that they have successfully been used in diagnosis (cf. Oksala and Therman, 1974). Although they obviously reflect something fundamental in the transformed cells, we do not know what causes them. Together with chromosome rearrangements, they provide the mechanism that creates the great variety of chromosome constitutions observed in cancer cells. These, in turn, are the material on which selection acts, promoting the fastest dividing cell types.

The Effect of Chromosome Changes on Tumor Development

Obviously the changes in chromosome number and structure cause tumor cells to become more and more different from the normal tissues of their origin. These changes provide the means for the stepwise malignization of a cancer. Indeed it is well known that a fairly benign tumor or leukemia may at one jump become much more malignant. Furthermore, a cancer that has responded satisfactorily to chemotherapy up to a point may flare up all at once, having become resistant to the treatment. On the other hand, when a leukemia undergoes remission, the chromosomally abnormal cells decrease in number or vanish altogether.

Cell Hybridization Studies

Cell hybridization, which is discussed in more detail in Chapter XIX has become one of the most important tools in somatic cell genetics. In cancer research, too, cell hybridization has served many purposes (cf. Miller, 1974). One of the most important achievements of cell hybridization is the mapping of various human genes on specific chromosomes. However, attempts to localize cancer-causing or cancer-promoting genes have so far not met with great success.

The localization of the tumor-inducing virus SV40 on chromosome 7 in human cells transformed in vitro has, however, been achieved with this method. Mouse-man hybrid cells in which the only human chromosome retained is 7 form malignant tumors, whereas cells without chromosome 7 are unable to do so. Hybrid cells containing this chromosome are also able to release SV40 viruses, whereas cells without it do not seem to include the virus either (cf. Croce and Koprowski, 1978).

Conclusions

At first glance, the evidence for the chromosomal origin of many tumors appears fairly convincing. Not only the close correlation between the ability of various agents to break chromosomes and to induce cancerous tumors, but also the observations on specific chromosome abnormalities in many types of malignancies seem to speak for this assumption.

On the other hand, the interesting observation that certain types of cancer cells undergo a series of apparently predetermined chromosome changes *after* they have become malignant seriously undermines the assumption that if the same abnormal chromosome constitution is characteristic of a specific type of tumor, it must be the cause of the malignant transformation.

The present state of knowledge can perhaps be best summarized in the words of Levan et al (1977, p. 28): "The original transformation may involve only submicroscopic changes and lead to premalignant and moderately malignant conditions. Subsequent chromosome changes—amplification (duplication) of the affected segments and/or loss of the homologous normal segments—underlie the progression towards states of increasing malignancy."

References

Atkin NB (1974) Chromosomes in human malignant tumors: a review and assessment. In: German J (ed) Chromosomes and cancer. Wiley, New York, pp 375–422

Bayreuther K (1952) Der Chromosomenbestand des Ehrlich-Ascites-Tumors der Maus. Naturforsch 7:554–557

Croce CM, Koprowski H (1978) The genetics of human cancer. Sci Am 238:117–125

DiPaolo JA, Popescu NC (1976) Relationship of chromosome changes to neoplastic cell transformation. Am J Pathol 85:709–738

Francke U (1976) Retinoblastoma and chromosome 13. Cytogenet Cell Genet 16:131–134

Grouchy J de, Turleau C (1974) Clonal evolution in the myeloid leukemias. In: German J (ed) Chromosomes and cancer. Wiley, New York, pp 287–311

Harnden DG (1974) Viruses, chromosomes, and tumors: the interaction between viruses and chromosomes. In: German J (ed) Chromosomes and cancer. New York, pp 151–190

Heneen WK (1976) HeLa cells and their possible contamination of other cell lines: karyotype studies. Hereditas 82:217–247

Lehto L (1963) Cytology of the human Fallopian tube. Acta Obstet Gynecol Scand Suppl 42:1–95

Levan A, Levan G, Mitelman F (1977) Chromosomes and cancer. Hereditas 86:15–30

Lynch HT, Harris RE, Organ CH Jr et al (1977) The surgeon, genetics, and cancer control: the cancer family syndrome. Ann Surg 185:435–440

Makino S (1975) Human chromosomes. Igaku Shoin, Tokyo

Makino, S (1974) Cytogenetics of canine venereal tumors: worldwide distribution and a common karyotype. In: German J (ed) Chromosomes and cancer. Wiley, New York, pp 335–372

Mark J (1974) The human meningioma: a benign tumor with specific chromosome characteristics. In: German J (ed) Chromosomes and cancer. Wiley, New York, pp 497–517

Miller OJ (1974) Cell hybridization in the study of the malignant process, including cytogenetic aspects. In: German J (ed) Chromosomes and cancer. Wiley, New York, pp 521–563

Miller RW (1967) Persons with exceptionally high risk of leukemia. Cancer Res 27:2420–2423

Mitelman F, Levan G (1978) Clustering of aberrations to specific chromosomes in human neoplasms. III. Incidence and geographic distribution of chromosome aberrations in 856 cases. Hereditas 89:207–232

Mitelman F, Levan G, Nilsson P et al (1976) Non-random karyotypic evolution in chronic myeloid leukemia. Int J Cancer 18:24–30

Müller H, Stalder GR (1976) Chromosomes and human neoplasms. Achievements using new staining techniques. Europ J Pediatr 128:1–13

Nowell PC (1976) The clonal evolution of tumor cell populations. Science 194:23–28

Nowell PC, Hungerford DA (1960) A minute chromosome in human chronic granulocytic leukemia. Science 132:1497

Oksala T, Therman E (1974) Mitotic abnormalities and cancer. In: German J (ed) Chromosomes and cancer. Wiley, New York, pp 239–263

Rowley JD (1977) A possible role for nonrandom chromosomal changes in human hematologic malignancies. In: De la Chapelle A, Sorsa M (eds) Chromosomes today 6. Elsevier North-Holland, Amsterdam pp 345–355

Rowley JD (1973) A new consistent chromosomal abnormality in chronic myelogenous leukemia identified by quinacrine fluorescence and Giemsa staining. Nature (Lond) 243:290–293

Sandberg AA (1979) The chromosomes in human cancer and leukemia. Elsevier North-Holland, New York

Sandberg AA, Sakurai M (1974) Chromosomes in the causation and progression of cancer and leukemia. In: Busch H (ed) The molecular biology of cancer. Academic, New York, pp 81–106

Shaw MW, Chen TR (1974) The application of banding techniques to tumor chromosomes. In: German J (ed) Chromosomes and cancer. Wiley, New York, pp 135–150

Therman E, Timonen S (1950) Multipolar spindles in human cancer cells. Hereditas 36:393–405

Wolf U (1974) Theodor Boveri and his book "On the problem of the origin of malignant tumors." In: German J (ed) Chromosomes and cancer. Wiley, New York, pp 3–20

Yosida TH (1975) Chromosomal alterations and development of experimental tumors. In: Handbuch der allgemeinen Pathologie VI. Springer, Heidelberg, pp 677–753

Zech L, Haglund U, Nilsson K et al (1976) Characteristic chromosomal abnormalities in biopsies and lymphoid-cell lines from patients with Burkitt and non-Burkitt lymphomas. Int J Cancer 17:47–56

XIX
Mapping of Human Chromosomes

Mapping of Human Chromosomes

Mapping genes on the chromosomes is one of the fastest growing disciplines in human genetics. Reports on new assignments of genes to particular chromosomes and books and reviews dealing with various aspects of chromosome mapping are appearing regularly (cf. Ruddle and Creagan, 1975; Creagan and Ruddle, 1977; McKusick and Ruddle, 1977; Shows, 1978). In addition, the developments in this field have been reviewed almost every year at international meetings: New Haven in 1973; Rotterdam, 1974; Baltimore, 1975; Winnipeg, 1977; and Edinburgh, 1979.

The considerable advances in gene mapping provide a good example of the successful application of different approaches to the same problem. These range from family studies through somatic cell genetics and from cytogenetics to biochemical and molecular studies. Often a gene assignment based on one technique has been confirmed with another. The traits for which loci have been mapped include inherited diseases, enzymes, serum and other proteins, surface and blood group antigens, RNA markers, susceptibility to drugs and toxins, viral markers, gene-regulating markers, and blood-clotting factors (cf. Shows, 1978).

In gene mapping as in many other branches of cytogenetics, chromosome banding provided a breakthrough. Genes can be assigned not only to individual chromosomes but also to specific chromosome regions.

Gene assignments are classified as *confirmed* when at least two studies have come to the same conclusion; *provisional* when only one determination exists; and *controversial* when different studies provide contradictory assignments (cf. Donald and Hamerton, 1978).

Family Studies

It is usually easy to decide, based on family studies, whether a gene lies on an autosome or on one of the sex chromosomes. An X-linked gene is never inherited from father to son, whereas a Y-linked gene is, but cannot be passed from father to daughter. The first human gene to be assigned to a specific chromosome was red-green color blindness, which was assigned to the X chromosome as early as 1911 (cf. McKusick and Ruddle, 1977), and now more than 100 X-linked genes are known (cf. McKusick, 1978). Three genes have been reported on the Y chromosome, but only the assignment of the one responsible for the H-Y antigen is firmly established (Chapter XII). Studies of linkage and crossing-over show whether or not two or more genes are on the same chromosome (*syntenic*), and what their relative distances are.

Marker Chromosomes

The assignment of a gene or a linkage group to a specific chromosome can be done on the basis of family studies only if a suitable marker chromosome is available. Typical markers consist of heterochromatic variants, fragile regions, or structurally abnormal chromosomes. The first gene assigned by means of a marker chromosome was the Duffy blood group locus, which segregated with a large variant of the centric heterochromatin in chromosome 1 (Donahue et al, 1968). The gene for α-haptoglobin was linked to a fragile site on 16q (such a fragile site is illustrated in Fig. VII.5 a and b) in 30 family members and was separated from it in three (Magenis et al, 1970). This finding indicates that the α-Hp locus lies near the fragile site on 16q. A gene causing mental retardation follows the fragile region near the distal end of Xq (cf. Howard-Peebles and Stoddard, 1979).

The major histocompatibility complex (HLA) was assigned to chromosome 6 by means of a pericentric inversion segregating in a family (Lamm et al, 1974). Reciprocal translocations involving chromosome 6 made the further localization of this gene to 6p21 possible (Breuning et al, 1977; Francke et al, 1977). This assignment was, in turn, confirmed through a study of partial trisomy for the segment distal to 6p21 that resulted from crossing-over in a pericentric inversion (Pearson et al, 1979).

Cell Hybridization

Although refinements in banding techniques have made more and more chromosome markers available (it is claimed that by now every human being can be distinguished by them), gene mapping owes its greatest advances to other methods, especially to cell hybridization. All cell hybridization techniques are based on the observation that somatic cells of the same species or of two different species fuse under certain conditions. For purposes of human chromosome mapping a human and a rodent cell are usually hybridized. Mouse cells are often used as the nonhuman parent strain. The following criteria should be considered when the cell types are chosen for human gene mapping (cf. Creagan and Ruddle, 1977):

1) The cells grow rapidly in culture.
2) The cells are easy to hybridize, and the hybrid cells divide in culture.
3) The chromosomes of the parent cells can be identified without difficulty (Fig. XIX.1).
4) Human chromosomes are unilaterally lost from the hybrid cells (see following paragraphs).
5) The human phenotypic markers can be easily determined and are distinguishable from those of the other parent cell.

Somatic cells sometimes fuse spontaneously, although this event is extremely rare. The incidence of cell fusions can be increased considerably by treating cultured cells with inactivated Sendai virus or a chemical agent, such as polyethylene glycol (cf. Ruddle and Kucherlapati, 1974).

Fig. XIX.1. Part of a metaphase plate from a man-mouse hybrid cell stained with G-11. Mouse chromosomes (aneuploid cell line) are dark with light centric regions, human chromosomes (h) are light with dark centric regions, and translocation chromosomes (t) between the two species are part dark, part light (courtesy of RI DeMars).

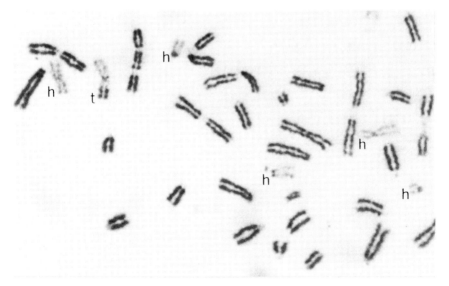

When two cells fuse, the hybrid has two nuclei at first; it is a *heterokaryon*. After nuclear fusion the cell is called a *synkaryon* (Fig. XIX.2).

Selection

Different selection systems play an important role in all cell hybridization methods. They include selection for a specific gene (or more accurately for the phenotype it causes), for hybrid cells against the background of parental cells, or for a specific chromosome or chromosome segment. Selection may be positive or negative, relative or absolute. The means used include culture media in which a certain cell type cannot grow or grows preferentially, or selective killing of one cell type with toxins, antibodies, or viruses to which the other cells are resistant.

Even an artificially increased cell fusion rate is so low that the hybrid cells have to be enriched relative to the parental cells (cf. Ruddle and Kucherlapati, 1974). Sometimes hybrid cells divide faster than the parent cells, but usually selection for the hybrids involves suppression of the parental cells. In the example in Fig. XIX.2, the mouse cells can be eliminated by a medium in which only cells that produce the enzyme E are able to grow. (The cell markers are often enzymes, which can be determined through their mobility in gel electrophoresis.) The human

Fig. XIX.2. Assignment of the gene coding for the enzyme E to a specific human chromosome (black) through fusion of a human cell (nucleus black) with a mouse cell (nucleus outlined) followed by cell selection.

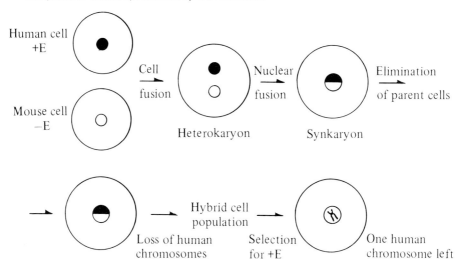

cells, on the other hand, can be killed with diphtheria toxin or the cardiac glycoside ouabain to which the mouse and hybrid cells are insensitive.

Figure XIX.2 illustrates the steps in the assignment of the gene coding for a hypothetical enzyme E to a specific human chromosome. After the fusion of the human and mouse cells and nuclei, the parent cells are eliminated. During the next phase the hybrid cells begin to lose human chromosomes in a random fashion. Cell hybridization can be regarded as a parasexual process in which meiotic segregation is replaced by random chromosome loss. Neither the reason for nor the exact mechanism of the process of chromosome elimination is understood, although it plays an important role in the gene mapping procedure.

Chromosome elimination leads to the formation of hybrid clones that differ in their human chromosome content. At this point, selection for or against a certain phenotype in the hybrid cells is an important tool in the gene mapping process. In our example (Fig. XIX.2), a selection system is used that allows only those cells producing enzyme E to survive. Naturally the clearest result is obtained when cells with phenotype E have only one human chromosome left, which must therefore be the site of gene E. The first human enzyme to be assigned in this way was thymidine kinase (TK) to chromosome 17 (cf. Ruddle and Creagan, 1975).

The usual result of random chromosome elimination is a collection of clones with different combinations of human chromosomes. A selected *clone panel* may enable the investigator to assign a gene to a specific chromosome. If all the clones displaying a particular phenotype have one and only one human chromosome in common, the conclusion is inevitable that this one is the site of the gene (cf. Ruddle and Creagan, 1975). With the clone panel technique it is also possible to determine whether or not two or more genes are syntenic.

Chromosome Translocations

The order of genes on a chromosome, as well as their assignment to a specific chromosome segment, can be determined by means of reciprocal translocations involving the relevant chromosome. The co-occurrence in a hybrid cell of a particular phenotype accompanied by one of the translocation chromosomes restricts the location of the gene to the part of the chromosome involved. By the use of several, partly overlapping translocations, the assignment of a gene can be limited to a smaller and smaller chromosome segment.

This was done for three X-linked genes by means of a translocation in which almost the whole Xq was attached to the distal end of 14q (Allderdice et al, 1978). Cell hybridization studies showed that the genes for HGPRT (hypoxanthine-guanine phosphoribosyl-transferase), PGK (phosphoglycerate kinase), and G6PD (glucose-6-phosphate dehydrogenase) segregated with the long translocation chromosome as did the autosomal gene NP (nucleoside phosphorylase). The three X-linked genes could thus be assigned to Xq and the autosomal gene to 14q (Ricciuti and Ruddle, 1973). By means of other X-autosomal translocations it was possible to demonstrate that the order of the three genes was: centromere-PGK-HGPRT-G6PD and to determine the limits of the segments within which each of them must be situated.

Two different gene maps have been developed for many human chromosomes. The one type of map contains the genes whose order and relative distances have been determined with family studies, whereas in the other type the genes that can be studied on the cellular level are assigned to specific segments (Fig. XIX.3) (cf. McKusick and Ruddle, 1977; Shows, 1978). Attempts have been made to bring the two maps together (cf. Francke et al, 1977).

Fig. XIX.3. Gene map of human chromosome 1, utilizing chromosome translocations and cell hybridization compared with another map of the same chromosome based on family studies (Shows, 1978).

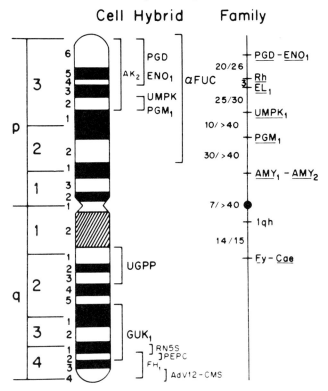

The enzyme TK, already mentioned, which had been assigned to chromosome 17, was found to lie on 17q when this arm was translocated to a mouse chromosome in a hybrid cell (cf. Ruddle and Kucherlapati, 1974). This was the first example of a translocation between two chromosomes of different species. An even more accurate assignment to the segment distal to 17q21 was possible with the help of a translocation between two human chromosomes (Yoshida and Matsuya, 1976).

Transfer of Microcells and Single Chromosomes

A refinement of the cell hybridization technique is the fusion of a "microcell" with a normal cell. A microcell consists of a nucleus with a few (or even a single) chromosomes surrounded by a small amount of cytoplasm. Microcells are prepared by treating normal cells with colchicine to scatter the chromosomes around the cell and then treating the cell with cytochalasin B to break it up. Such a microcell can then be fused in the usual way with a complete recipient cell (cf. Fournier and Ruddle, 1977).

Another modification of the cell hybridization techniques is the transfer of a single metaphase chromosome into a recipient cell. The transferred chromosome falls into fragments in the alien cytoplasm, leaving the gene under study attached to a segment of DNA. Such hybrid cells are unstable until, many cell generations later, the fragment may attach itself to a host chromosome.

Gene Dosage

Mapping by means of gene dosage is either qualitative or quantitative. In the former case the number of different gene products is compared with the number of genes present. Quantitative studies, on the other hand, correlate the number of alleles with the quantity of a gene product (cf. Creagan and Ruddle, 1977). Both types of studies take advantage of chromosome deletions and duplications (partial trisomy or monosomy). No primary gene assignment has been made solely by means of trisomy mapping, but a number of previous assignments has been confirmed. A deletion of the distal end of 2p enabled the mapping of the gene for acid phosphatase to the missing segment (cf. Aitken et al, 1976). Several other assignments have been confirmed with deletion mapping (cf. McKusick and Ruddle, 1977).

Exclusion mapping is the reverse of trisomy and monosomy mapping. This method determines the genes that are *not* affected by the duplication or deletion of a chromosome segment. A study of 20 deletions made possible the exclusion of an average of four genes from each deleted segment (cf. Aitken et al, 1976).

In-situ Hybridization

When a probe of DNA or complementary RNA is labeled with a radioactive isotope and hybridized to a chromosome preparation, the original site of the DNA is marked with silver grains after autoradiography (cf. Pardue and Gall, 1975). In this way the human satellite DNAs have been localized to the centric and Y heterochromatin (cf. Yunis et al, 1977). The repeated genes, which occur in 100 to 200 copies in the human chromosome complement, have also been mapped by direct hybridization (cf. Evans and Atwood, 1978). They include the genes coding for the ribosomal proteins 18S and 28S, which are situated on the satellite stalks of acrocentric chromosomes. The genes producing the 5S ribosomal protein have been assigned to 1q42-q43 (cf. Steffensen, 1977). The assignment of the histone genes, repeated about 40 times, to 7q32-q36 was confirmed by direct hybridization on prophase-banded chromosomes (Chandler et al, 1979).

Conclusions

Constitutive heterochromatin is naturally devoid of any mendelian genes, but there is convincing evidence that the Q-bright chromosome bands also consist to a considerable extent of intercalary heterochromatin and therefore contain relatively few genes. In contrast, the Q-dark regions have a high gene density (Chapter VI).

The human genome has been guessed to contain some 50,000 genes. However, only a fraction of them is known. McKusick (1978 and unpublished) lists 1367 autosomal genes whose mode of inheritance is established and 1426 others for which it is tentative. Because of the easy detection of X-linkage, as many as 113 genes are known on this chromosome, whereas 108 others are provisionally assigned to it. Of the autosomal genes, more than 200 have been mapped more or less firmly to specific chromosomes (McKusick, 1978 and unpublished).

The unambiguous assignment of single, unique genes has so far not been possible with direct hybridization, although, as mentioned earlier, this has been achieved for the repeated genes. One of the limitations is the difficulty in obtaining the necessary amount of a probe that is specific and "hot" enough. Another obstacle is the resolving power of autoradiography. However, when these techniques are improved—as they undoubtedly will be—direct hybridization will be the preferred gene mapping method of the future (cf. Steffensen, 1977).

One of the goals of human genetics is the development of a complete gene map. This accomplishment lies far in the future, since the overwhelming majority of the genes is still unknown. One of the tasks ahead is the coordination of the results of formal genetics, somatic cell genetics, and molecular studies on chromosome structure. The crowning achievement on the molecular level will be the base sequencing of the total DNA of the human genome.

References

Aitken DA, Ferguson-Smith MA, Dick HM (1976) Gene mapping by exclusion: the current status. Cytogenet Cell Genet 16:256–265

Allderdice PW, Miller OJ, Miller DA et al (1978) Spreading of inactivation in an (X;14) translocation. Am J Med Genet 2:233–240

Breuning MH, Berg-Loonen EM van den, Bernini LF et al (1977) Localization of HLA on the short arm of chromosome 6. Hum Genet 37:131–139

Chandler ME, Kedes LH, Cohn RH (1979) Genes coding for histone proteins in man are located on the distal end of the long arm of chromosome 7. Science 205:908–910

Creagan RP, Ruddle FH (1977) New approaches to human gene mapping by somatic cell genetics. In: Yunis JJ (ed) Molecular structure of human chromosomes. Academic, New York, pp 89–142

Donahue RP, Bias WB, Renwick JM et al (1968) Probable assignment of the Duffy blood group locus to chromosome 1 in man. Proc Natl Acad Sci USA 61:949–955

Donald LJ, Hamerton JL (1978) A summary of the human gene map, 1973–1977. Cytogenet Cell Genet 22:5–11

Evans HJ, Atwood KC (1978) Report of the committee on in situ hybridization. Cytogenet Cell Genet 22:146–149

Fournier REK, Ruddle FH (1977) Microcell-mediated chromosome transfer. In: Sparkes RS, Comings DE, Fox CF (eds). Molecular human cytogenetics. Academic, New York, pp 189–199

Francke U, George DL, Pellegrino MA (1977) Regional mapping of gene loci on human chromosomes 1 and 6 by interspecific hybridization of cells with a t(1;6)(p3200;p2100) translocation and by correlation with linkage data. In: Sparkes RS, Comings DE, Fox CF (eds). Molecular human cytogenetics. Academic, New York, pp 201–216

Howard-Peebles PN, Stoddard GR (1979) X-linked mental retardation with macro-orchidism and marker X chromosomes. Hum Genet 50:247–251

Lamm LU, Friedrich U, Petersen GB et al (1974) Assignment of the major histocompatibility complex to chromosome no. 6 in a family with a pericentric inversion. Hum Hered 24:273–284

Magenis RE, Hecht F, Lovrien EW (1970) Heritable fragile site on chromosome 16: probable localization of haptoglobin locus in man. Science 170:85–87

McKusick VA (1978) Mendelian inheritance in man, 5th edn. Johns Hopkins University Press, Baltimore

McKusick VA, Ruddle FH (1977) The status of the gene map of the human chromosomes. Science 196:390–405

Pardue ML, Gall JG (1975) Nucleic acid hybridization to the DNA of cytological preparations. In: Prescott DM (ed) Methods in cell biology, Vol 10. Academic, New York, pp 1–16

Pearson G, Mann JD, Bensen J et al (1979) Inversion duplication of chromosome 6 with trisomic codominant expression of HLA antigens. Am J Hum Genet 31:29–34

Ricciuti FC, Ruddle FH (1973) Assignment of three gene loci (PGK, HGPRT, G6PD) to the long arm of the human X chromosome by somatic cell genetics. Genetics 74:661–678

Ruddle FH, Creagan RP (1975) Parasexual approaches to the genetics of man. In: Roman HL, Campbell A, Sandler LM (eds). Annual review of genetics, Vol 9. Annual Reviews, Palo Alto, pp 407–486

Ruddle FH, Kucherlapati RS (1974) Hybrid cells and human genes. Sci Am 231:36–44

Shows TB (1978) Mapping the human genome and metabolic diseases. Birth Defects, Proc 5th Internat Conf: 66–84

Steffensen DM (1977) Human gene localization by RNA:DNA hybridization *in situ*. In: Yunis JJ (ed) Molecular structure of human chromosomes. Academic, New York, pp 59–88

Yoshida MC, Matsuya Y (1976) Confirmation of the human thymidine kinase locus, 17q21→17qter, by means of a man-mouse somatic cell hybrid, D98/AH-2 X LMTK⁻ C1-1D. Hum Genet 31:235–239

Yunis JJ, Tsai MY, Willey AM (1977) Molecular organization and function of the human genome. In: Yunis JJ (ed) Molecular structure of human chromosomes. Academic, New York, pp 1–33

Author Index

Subject Index